Forest
Sustainability

FOREST

HISTORY

SOCIETY

ISSUES

SERIES

Forest Sustainability
The History, the Challenge, the Promise

Donald W. Floyd

The Forest History Society
701 Vickers Avenue
Durham, North Carolina 27701
919 682-9319
www.lib.duke.edu/forest

First edition

The Forest History Society is a nonprofit, educational institution dedicated to the advancement of historical understanding of human interaction with the forest environment. The Society was established in 1946. Interpretations and conclusions in FHS publications are those of the authors; the Society takes responsibility for the selection of topics, the competence of the authors, and their freedom of inquiry.

Publication of this book was supported by grants to the FHS from the Laird Norton Endowment Foundation, Weyerhaeuser Company Foundation, and the USDA Forest Service–State and Private Forestry and International Programs.

The author acknowledges the contributions of Sarah Vonhof and Heather Seyfang to the sections on the etymology of *sustainability* and the case studies.

Cover: Düsterer Tag ("Gloomy Day"), 1565, by Pieter Bruegel the Elder, is one of the earliest European landscape paintings. Peasants are cutting and gathering tree limbs for firewood. KUNSTHISTORISCHES MUSEUM, VIENNA.

Library of Congress Cataloging-in-Publication Data
Floyd, Donald W., 1951-
  Forest sustainability : the history, the challenge, the promise /
Donald W. Floyd.— 1st ed.
    p. cm. — (Forest History Society issues series)
  Includes bibliographical references (p.   ).
  ISBN 0-89030-061-5 (pbk.)
  1. Sustainable forestry. 2. Forest policy. I. Forest History
Society. II. Title. III. Series.
  SD387.S87 F58 2002
  634.9'2—dc21
                              2001007187

Contents

Tables and Figures

FOREWORD

Within the discussion on sustainability one finds many perspectives. And nearly all of them are based upon rather singular views, whether individual or institutional. In this book, Donald W. Floyd challenges the reader to explore myriad considerations of the origins, development, and current context of sustainable forestry.

Throughout the book, exemplars are described against which current issues may be compared. The value of collecting and monitoring information is made abundantly clear. Without a historical context, how can we objectively evaluate our progress? The Sustainable Forestry Partnership strongly believes in the value of such case studies. However different in scale and time, our experiences provide the basis for social learning. Throughout this book, readers will find experiences that can be assembled to inform and enrich their own decision-making environments.

We believe this book makes several fundamental contributions. First, the articulation of principles surrounding sustainability allows us to apply and test the concept against time and contemporary situations. Such principles can be used as cornerstones against which progress toward sustainability might be assessed. Second, the importance of context is illustrated through descriptions of historical events with commensurate lessons. To the extent we are a learning society, documenting these lessons increases the capital stock of social knowledge and provides the ability to incrementally improve their application. Third, the need for social engagement to more fully involve people in making decisions is made clear. As Floyd says, "problems of sustainability cannot be solved by science alone." It is both intuitively and empirically obvious that our collective future is enhanced through our collective involvement.

Where to from here? Forests sustain life, and the trajectory of their sustainability, and thus ours, will be largely defined by our ability to learn. Participation is the key, yet the social dimensions of the sustainability triangle are perhaps least understood. One thing is clear: the conversation itself must be sustained.

—*A. Scott Reed and Michael P. Washburn*
The Sustainable Forestry Partnership

OVERVIEW

This short book is a primer on a complicated subject. It introduces forest sustainability and places it in its historical context. Forest sustainability is a concept that has evolved over thousands of generations and only gradually emerged as we now recognize it. It begins with the forest as a source of the sacred and gradually expands over 5,000 years to include the forest as source of material wealth and eventually as an essential site for life-sustaining processes. Here are some important issues to consider:

• Problems of food production and forest protection are universal in human culture. We have been converting forests to other uses (principally agriculture and urban areas) and occasionally creating new forests for at least 10,000 years. Understanding the history of this process and our responses to it is critical for those who would shape the legacies we leave.

• "Sustainable forestry" relies on a set of technologies, management practices, and social and economic institutions that communities, governments, and industry use to ensure the long-term viability of their forests. The values and choices that these groups select lead to different outcomes. Because many of the products and benefits that we seek to sustain in the forest are mutually exclusive at the local level, sustaining the forest requires difficult choices and public consensus.

• The things that people want from their forests change over time. This challenges our concept of sustainability and our prospects for achieving it. In the past 5,000 years forests have been a place for worship, a source of fuel and fiber, a reservoir for biodiversity, and an important contributor to global climate processes. It is difficult to predict the things that future generations will want from their forests. Sustainability seeks a way to accommodate our expanding desires.

• The role of intensive forest management is an important issue in meeting market demands while protecting environmentally sensitive natural forests. By intensively managing some forests for fiber and fuel and managing other forests to maximize biodiversity, wildlife habitat, or watershed values, sustainable forest management attempts to achieve a combination of uses that maintain the forest as well as the human communities that depend on it.

• In the developing nations, particularly in the Southern Hemisphere, the principal challenge to sustaining the forest is preventing its permanent conversion to other land uses, especially agriculture. This problem is directly linked to human population growth.

• Permanent forest conversion is a problem near rapidly growing cities in the Northern and Southern Hemispheres. In much of the Northern Hemisphere, total forest area is increasing as food production is concentrated on smaller areas. Forest expansion is largely the result of switching from fuelwood to fossil fuels. Fuel conversion and new agricultural technologies allowed the forests to regrow after reaching a low point in the early 20th century. In the developed nations, concern about sustainable forestry focuses on the ways that humans change native forests and create new forests.

• The many definitions of *forest* in inventory methods worldwide hamper our ability to assess deforestation and hence forest sustainability. Compatible forest inventories and a congruent set of criteria and indicators are a prerequisite to defining, assessing, and achieving forest sustainability.

• Forest sustainability on a global basis is a distant, worthy, and perhaps unobtainable goal without significant changes in technology and human population control. Nevertheless, it remains a goal that we must seek, just as we strive for "perfect justice" and "absolute truth."

In the next ten years many forests will be destroyed, some species will be eradicated, other forests will be created, and all our forests will change in subtle and not-so-subtle ways. Human population will continue to grow and the forests that each of us have known will be different. In the end, what we will truly sustain is change—and the knowledge that we have the power to shape our descendants' future.

INTRODUCTION

Concern for the condition of the world's forests is widespread. News reports highlight the destruction of tropical forests. Demonstrators protest international trade in forest products. Conservationists bemoan the loss of biodiversity. Multinational corporations seek certification that their forests are managed sustainably. Governments enter into international agreements that promise sustainable management of their forests.

That global concern for forest protection and environmental conservation increased in the 1980s and 1990s should not be cause for surprise. Human population reached an estimated 6 billion in 1999. As the earth lost about 12 million hectares (more than 29 million acres) of forest each year in the 1980s and 1990s, threats to endangered species and biodiversity became apparent. The 1990s left many of us with a growing realization that the choices of future generations will be limited because of our inability to protect valuable ecosystems. Moreover, the 1990s were a period of unprecedented economic expansion for many North Americans and Europeans. The boom created enough wealth that many citizens in developed countries could afford to consider the prosperity of future generations.

Although popular concern is growing and seemingly disparate interests now appear to agree on it as a goal, there are serious obstacles to achieving sustainability. Living sustainably requires that we secure the needs and values of our own generation without compromising the ability of future generations to meet their needs and values. At a minimum, that means adequately feeding the current human population while simultaneously protecting the productive capacity of the earth.

But a quick review reveals two disquieting facts. The number of malnourished humans is greater now than it ever has been. Given today's economic systems and technologies, producing more food for tomorrow will most likely mean clearing more forestland for food production.

1

On a global basis, deforestation for agricultural production remains the most serious threat to the world's forests.

Second, humans have always altered the earth's environment to favor their own survival at the expense of other species, and undoubtedly we will continue to do so. Even if we can manage the world's forests to produce the water, fuel, paper, lumber, wildlife habitat, recreational opportunities, and carbon storage that we desire, forests managed for these products will not produce the same mix of values as unmanaged forests. Many of the forests we sustain will likely be very different from the ones we know.

The idea of sustainability is surprisingly simple: Resource consumption cannot exceed resource production over time. Sustainability is related to the wildlife biologist's concept of carrying capacity—the number of individuals that a habitat can support without damaging its productive capability. *Human* carrying capacity is not an absolute number, however. It is subject to a host of decisions about the quality of life. The carrying capacity of a planet where humans eat a nutritionally complete diet of yeast and algae is much greater than the carrying capacity of a planet where humans value songbirds and eat grain. Much of the difference (and thus our conception of sustainability) has to do with how and what humans choose.

As democracy and free markets replace centralized governments and economies throughout the world, it is only natural that people in every nation, once empowered to make individual choices, will choose houses over hovels, rich diets over hunger, and comfort over cold. Concern for achieving sustainability recognizes the limits imposed by both the laws of physics and human institutions.

Sustainability assumes that the choices of the current generation will not preclude the options of future generations. Because many of the resources we use are not renewable, this is impossible. A more realistic approach to sustainability suggests that we carefully consider our current choices so that future generations will have more options than they might otherwise have.

Sustainable forestry, sustainable agriculture, and human population are intertwined. Human population growth requires more food, and turning forestland into cropland is often the easiest way to expand food supplies. We have reached a point where additional loss of forests can affect water quality, biodiversity, and climate patterns. Such changes were apparent before on a regional scale, but now they occur a global scale.

Will new agricultural technologies be enough to slow the rate of deforestation? By substituting fossil fuels for animal agriculture and using pesticides, fertilizers, and genetically superior crop varieties, many nations in the Northern Hemisphere have reversed deforestation trends. Whether such a reversal is likely in the developing nations of the Southern Hemisphere remains to be seen. Without changes in human population growth and our collective ability to protect and manage natural resources, perhaps the best we can hope for is a system of environmental triage that minimizes ecological losses and relies on new technologies that convert natural resources to meet human needs more efficiently.

IN THE EYE
OF THE BEHOLDER

Trying to define sustainability and sustainable forestry is like trying to define "justice" or "democracy." There are many definitions and some consensus, but agreement over the specifics is elusive. If sustainability cannot be specifically defined, does that mean it is of little value? Foresters know there are many useful yet ambiguous terms, like "multiple use," "forest health," and "ecosystem." We come to grips with any new idea through discussion and debate, and we are still in the process of debating and defining the meanings of sustainability.

The most widely quoted definition of sustainability comes from a United Nations World Commission on Environment and Development publication. Known as the Bruntland Report after its director, Gro Harlem Bruntland, it appeared in 1987 under the title *Our Common Future* (Bruntland 1987).

The commission defined sustainable development as

...development that meets the needs of the present without compromising the ability of future generations to meet their own needs.

The implications of that seemingly simple definition are enormous. To achieve sustainability, we must first determine the needs of the current generation and the current rates of natural resource production and consumption, then make a reasonable estimate of the needs of an unspecified number of future generations. And predicting how future generations will value forests is even more challenging because the things that people want from their forests change over time.

One way to gain an understanding of sustainability is to examine the word itself. The word *sustain* means (1) to keep in existence or to maintain; (2) to supply with necessities or nourishment, provide for. The root is *tenere*, Latin for "to hold"; the prefix is derived from *sub*, "under, from below." The Latin compound *sustinere* and its English cognate, *sustain,* thus mean to hold up from below, to support, to last or endure.

Forest sustainability involves both ends and means—sustainable forests are the desired end and sustainable forest management is the means by which that end is achieved. The definition of *sustainability* in the *Dictionary of Forestry* (Helms 1998) is really a definition of a *sustainable forest:*

> ...the capacity of forests, ranging from stands to ecoregions, to maintain their health, productivity, diversity, and overall integrity, in the long run, in the context of human activity and use.

The subsequent definition for *sustainable forestry (sustainable forest management)* is one of the longest entries in the *Dictionary of Forestry.* It is adapted from the Montreal Protocol, an international convention adopted by many nations in the Western Hemisphere. It is worthwhile examining it:

> *[T]his evolving concept has several definitions* 1. The practice of meeting the forest resource needs and values of the present without compromising the similar capability of future generations —*note* sustainable forest management involves practicing a land stewardship ethic that integrates the reforestation, managing, growing, nurturing, and harvesting of trees for useful products with the conservation of soil, air and water quality, wildlife and fish habitat, and aesthetics.
>
> 2. The stewardship and use of forests and forest lands in a way, and at a rate, that maintains their biodiversity, productivity, regeneration capacity, vitality, and potential to fulfill, now and in the future, relevant ecological, economic, and social functions at local, national, and global levels, and that does not cause damage to other ecosystems —*note* criteria for sustainable forestry include (a) conservation of biological diversity, (b) maintenance of productive capacity of forest ecosystems, (c) maintenance of forest ecosystem health and vitality, (d) conservation and maintenance of soil and water resources, (e) maintenance of forest contributions to global carbon cycles, (f) maintenance and enhancement of long-term multiple socioeconomic benefits to meet the needs of societies, and (g) a legal, institutional, and economic framework for forest conservation and sustainable management.

Sustainable forest management is related to but different from sustained yield—the amount of wood that a forest can produce continually. The concept of sustained yield, which dates to the late Middle Ages in Central Europe, was gradually expanded to include the perpetual production of other forest outputs, including water, recreation, fish and wildlife habitat, and livestock forage. Yet sustainable forestry implies managing the forest for more than outputs; it focuses on processes.

Sustainable forestry is closely related to the concepts of conservation and stewardship. The three terms all imply the use of resources.

Originally, *conserve* meant to protect or preserve. Its Latin root is *servare*, "to keep, to watch, to maintain." A conservatory, for example, was a preservative for food and for medicine. In historical uses, then, conservation was associated with setting aside and preserving, and it connoted a consideration for future need.

Sustained yield—the amount of wood that a forest can produce in perpetuity—differs from sustainable forestry, which seeks to sustain communities, economies, and all the elements of a forest, in addition to timber supply. PHOTO COURTESY OF STEVEN ANDERSON.

The word *steward* derives from the 10th-century Old English *stigweard:* guard *(weard)* of the hall *(stig),* and thus a manager of a household or estate. Stewardship implies husbandry and the responsibility to manage something for someone else. Stewardship has always been associated with responsibility, and it connotes keeping and maintaining.

To summarize, *sustain* means to hold up, support, endure; *conserve* means to preserve and protect; and *stewardship* is the responsibility to manage and maintain. Thus, *sustainability*, *conservation*, and *stewardship* differ in nuance and coexist in current discussions about natural resources.

The Overlapping Ideas

In our context, sustainability reflects the broad range of values that individuals and societies seek to emphasize about time, human nature, and forests. The concept is easily stated but not easily implemented because the definitions vary with the values of the beholders. Forest industries emphasize ideas that ensure their profitability. Landowners in developing countries may emphasize agroforestry or community fuelwood production. Conservation groups usually emphasize non-market factors like biodiversity, ecosystem function, and protection of indigenous cultures. Like a mirror, we use sustainability to reflect the things that are most important to each of us.

In theory, because of the reproductive capability of living things, any renewable natural resource—trees, wildlife, parks, watersheds—can be managed for continuing productivity over hundreds or even thousands of years. The critical factor is that the amount of product harvested cannot exceed the periodic production of the system. But even this is complicated, because when one product or mix of products is favored, other elements of the system may decline.

Each interest selects the mix of values and kind of forest that best meet its needs, but there seems to be general agreement that forest sustainability comprises three elements:

• Ecological sustainability: maintenance of biological diversity and the integrity of ecological processes and systems.

• Social sustainability: maintenance of the human community that depends upon the forest.

• Economic sustainability: maintenance of companies, communities, and families that are economically dependent on forests.

Ecological sustainability. In the 1970s and 1980s biologists began to discuss how they might manage very large areas of land to conserve biodiversity. This idea, which eventually became known as ecosystem management, emphasized maintaining the ecological functions of the landscape rather than producing products. This was a subtle but very important change in management philosophy; it suggested that managers should first ensure the integrity of ecological processes and then consider the availability of forest products. Ecosystem management was controversial among forest managers who had been trained to emphasize the perpetual production of timber, but it was widely supported by ecologists and wildlife biologists and was eventually adopted as policy on most public lands in the United States and many other nations in the 1990s.

Like ecosystem management, ecological sustainability emphasizes the stability and resilience of biological and physical systems. Ecological sustainability requires that consumers, producers, and decomposers exist in balance over time, processing the energy from solar radiation and more limited geothermal energy to maintain life and transform matter. Ecological sustainability implies maintenance of biological diversity,

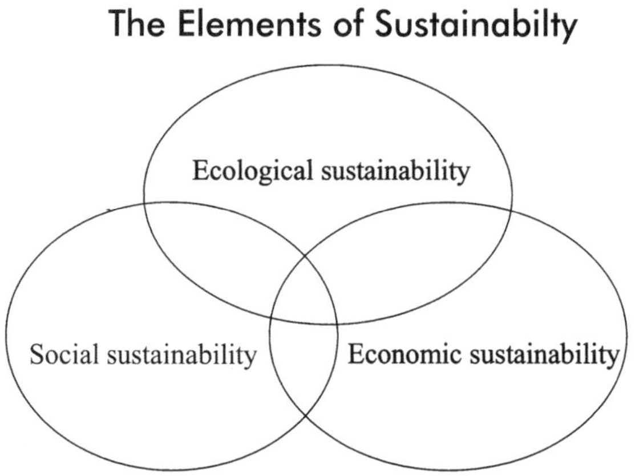

The Elements of Sustainabilty

Figure 1. Conceptually, forest sustainability is the area occupied by the overlap among the three elements. In other words, to be sustainable, forest management must be ecologically, socially, and economically sound.

and on this point there is considerable debate. Some ecologists argue that the ability of ecosystems to respond to disturbance is enhanced by the presence of many species. But how many and which species must be sustained to keep a system resilient?

In the United States, we have tried to address this issue by creating a system of parks and reserves where species and habitat preservation is given priority. In addition, the Endangered Species Act gives the federal government power to protect threatened and endangered populations on private as well as public lands. There have been both successes and failures with this approach. The conflicts are significant and the answers are not yet known.

Many writers argue for conservation of biodiversity because genetic material may prove useful for solving human health problems; they recall Leopold's (1949) first law of intelligent tinkering: to keep all the parts. Some argue that maintenance of biodiversity has a moral component and should be the focal point of public policy; others suggest there is not yet enough evidence for recommending policy changes.

Finally, there are the ecological processes or "services" that forests provide—in energy flow, nutrient cycling, and the hydrologic cycle. Using forests to capture excess carbon from fossil fuel consumption is another example. Many believe that forests can moderate the effects of human-induced global climate change and offset some of the carbon emissions produced by industrial nations.

Social sustainability. The social concept of sustainability emphasizes the resilience of human social and cultural systems. Aspects include raising living standards, ensuring the well-being of future generations, promoting public participation in decision making, and protecting indigenous knowledge. Like the moral aspects of the biodiversity argument, these values depend on one's preferred outcome. They are rooted in the modern notion of human rights, including the rights of self-determination and cultural preservation of indigenous people. Many international development organizations and human rights groups have interests in sustainability because of their commitment to social development.

Sustainable forestry in the developing world is closely tied to the sociocultural conception of sustainable development. Its political ideology is apparent in the social and economic criteria and indicators sometimes used to assess forest management.

Economic sustainability. Economists suggest that sustainable development (and by extension sustainable forestry) is principally concerned with managing tradeoffs among economic, sociocultural, and environmental objectives. For many economists, sustainability means maximizing the flow of income that can be generated while at least maintaining the stock of assets that yield those benefits.

Like managing a goose that lays golden eggs, economic sustainability requires maintaining our natural capital—the physical and biological processes that support life. We understand that some of this capital (coal, for example) is nonrenewable, so we must either limit our use of these resources to preserve the options of future generations or find new technologies. If we use nonrenewable resources to meet current needs, economic sustainability suggests that we should compensate future generations for the resources that will no longer be available to them.

Integrating the Perspectives

Each of those three approaches emphasizes different aspects of sustainability, and each interest interprets the concept through the lens of its own paradigm. The ecologists emphasize the interrelationships of the biophysical systems. Citizen activists and social scientists—political scientists, sociologists, anthropologists—focus on how humans as individuals and groups make decisions about resource use and conservation. Some economists and forest industries tend to emphasize tradeoffs and efficiency. Ultimately the different approaches must be integrated. Sustainability suggests that all the approaches are important and that all the values must be conserved, but we do not yet have effective mechanisms for allocating the competing values.

Many interest groups begin by acknowledging the importance of all forest values. Because they prefer certain values (e.g., global ecological services, protection of indigenous rights, biodiversity, timber or water production), however, they suggest that once some threshold amount of their favored value is maintained, then other values and products may be harvested to the extent that the primary value is not diminished. This produces many different versions of sustainability.

We are left with two questions: Who will decide how to meet our needs, and which methods will be used to evaluate the choices? Like Plato and his philosopher kings, some people want forest ecologist kings and queens to make the allocations. Others will argue for market-based decisions that emphasize efficient individual choice. Perhaps the answers will come from consensus derived from collective process, but such consensus has been rarely achieved.

FROM SACRED TO SUSTAINABLE

Sustainability is an idea that has evolved over several thousand years. As with the history of any idea, there is no neat, linear chronology. Forest protection began with the nearly universal idea of nature as sacred, a place where gods resided. As population pressures increased, the once-sacred forest was cleared for agriculture and exploited for its material products. With the rise of conservation in Western Europe and several other cultures, however, came the more or less simultaneous rediscovery of nature as a place of moral instruction.

What distinguishes the modern concept of sustainability from its precursors—nature as sacred, nature as inspiration—is its broader focus on maintaining biological systems and human cultures and economies as well as the ability of the earth to benefit future generations. Sustainability reflects the gradual expansion of human ethical, spatial, and temporal concerns.

Nature as Sacred

From ancient tree worshippers to the modern tree sitters protesting logging in the redwood forests of northern California, the idea that nature is sacred resonates in many cultures. It is one of the dominant themes in American and European natural resource and environmental policy during the 19th and 20th centuries. The tension between protecting spiritual values in nature and managing natural resources to meet material needs is one factor that shapes the debate over sustainability.

In his 1890 classic, *The Golden Bough*, Sir James George Frazer documents the worship of trees as one of the oldest forms of religion. He finds examples throughout Europe, North America, Africa, Asia, and Australia. In ancient cultures tree spirits nurtured human life and were thought to bring rain, sun, and fertility. Protecting sacred trees by setting them off in reserves was common in northern Europe as well as ancient Greece, Rome, and India.

Several advanced societies evolved in Sumeria. Although the inhabitants of Uruk and Ur were able to build impressive temples like this ziggurat, eventually deforestation and salinization reduced food production and the society became unsustainable. PHOTO COURTESY OF HIRMER FOTOARCHIV.

One of the earliest and best examples of the struggle between the sacred and the material comes from ancient Sumeria. A rich agricultural society dependent on snowmelt carried by the Tigris and Euphrates Rivers began to develop here between 2000 and 3000 B.C. In the *Epic of Gilgamesh* we read that an early lord of the city of Uruk decided to build an even greater city. A vast forest grew nearby, but it was protected by the gods. Enlil, the chief Sumerian deity, appointed the demigod Humbaba to protect the forest and its creatures. As Gilgamesh began to clear the forest, a fight ensued. Gilgamesh slew Humbaba but the food and water he depended on were cursed.

The epic apparently describes actual deforestation that led to increasing salinization and siltation of agricultural lands. In about 3500 B.C. the Sumerians grew wheat and barley in roughly equal proportions. But wheat has little tolerance for salt. By about 2500 B.C. wheat constituted only 15 percent of the cereal crop. By 1700 B.C. it was no longer grown in southern Sumeria. By the time the Babylonians conquered the region in about 1800 B.C., crop yields were one-third what they had been in the time of Gilgamesh. After more than 1,000 years, deforestation had caused environmental changes that effectively ended food production. The epic illustrates the importance of protecting the sacred forest from those who would use it unwisely, and it has parallels in many other cultures.

In ancient Greece, sacred areas ranging from a few trees to thousands of acres were set aside for religious reverence because they were occupied by the gods. Worship took place in outdoor settings. Laws protecting the sacred groves—as sources of fuelwood and construction material as well as for sacred values—were enforced by religious and civil sanctions, including steep fines and ritual curses. The groves were not unlike our national parks—the last vestiges of pristine nature in otherwise developed areas. But as wood fuel became more scarce in the fourth century B.C., villagers depleted the resource, and protection of the sacred gave way to the material needs of the community.

In both the Sumerian and the ancient Greek examples, the protected trees were associated with the gods. Thus the earliest attempts at protecting the forest were associated with the idea that the forest was home to the sacred. The idea was current in ancient Rome as well, where despite general deforestation, sacred groves near Rome survived. There is even evidence that planted groves sometimes took on the aspect of managed tree plantations, with both fruit-bearing trees and species valued chiefly for their timber. But the decline of the Roman Empire and the increasing influence of Christianity eventually led to the destruction of the ancient sacred groves in Europe in favor of economic uses.

Modern environmental scholars have argued both sides of the questions about stewardship of natural resources in the Judeo-Christian tradition. Restrictions on the harvest of game in the Mosaic law may be one of the earliest written examples of concern for conservation:

> If a bird's nest chance to be before thee in the way, in any tree or on the ground, with young ones or eggs, and the dam sitting upon the young or upon the eggs, thou shalt not take the dam with the young: thou shalt in any wise let the dam go, but the young thou mayest take unto thyself; that it may be well with thee and that thou mayest prolong thy day (Deuteronomy 22:6).

In a widely reprinted 1967 article, Lynn White argued that Christian dogma, which seems to separate humans from nature, was a principal cause of our current environmental dilemma. In fact, both themes—protection of the sacred and conservation of the material—are common in the Bible. The analogy of Earth as a garden and humans as gardeners recurs throughout the Old and New Testaments.

The Church taught that nature was placed by God in the hands of humans, who were to care for it as God's stewards. But monotheism also taught that God was separate from His creation and denied any inherent sacredness in nature.

Some of the deforestation of Western Europe before 1000 A.D. can be attributed to the Church. Separating themselves from worldly sins and temptations, early Christian monks established monasteries deep in the wild forests of Europe, western Asia, and the mountains of North Africa. Their desire to destroy the sacred groves of the pagans and humble themselves before God through manual labor brought lasting changes to the landscape.

Trees and forests have played a central role in many of the world's religions for thousands of years. These monks are praying at the Bodhi tree in Bodh Gaya, India—the site of the Buddha's enlightenment. PHOTO COURTESY OF SACREDSITES.COM AND MARTIN GRAY.

Well before 1000 A.D., the effect of forest clearing by the monastic orders was apparent. The invention of an efficient horse collar made it easier to clear the forest, drain wetlands, and move heavy objects long distances. Saws began to supplement the axe. As new technologies, like wind and water power, were harnessed for land clearing, food production, and sawing, the pace of deforestation quickened. The sacred forests of Europe, North Africa, and western Asia had nearly vanished.

Sacred groves remain today in India and in many areas where indigenous cultures continue to hold sway. But they are disappearing rapidly as the modern, globalized economy transforms the isolated remainders.

The Deforestation-Decline-Reforestation Cycle

For much of our history, humans lived in small bands, migrating according to the seasonal availability of food and congregating for ceremonial and social activities when food supplies permitted. Because there were relatively few people and the technologies they employed were primitive, their way of life was "sustainable." Until the advent of agriculture, the extent of the forest had more to do with climate change than with human effects.

Beginning about 10,000 years ago, agriculture introduced profound changes in ecological systems as well as in human culture. As humans diverted biological and energy resources to their own use—first in southwestern Asia, and slightly later in China and Mesoamerica—their populations grew and reshaped the environment.

Industrialization followed, and humans sought energy for firing ceramics, smelting bronze, and other energy-intensive processes. The energy came from wood, and deforestation was the by-product. Over the past 4,000 years, since the introduction of metallurgy, forested regions have undergone repeated cycles of deforestation, sometimes followed by regrowth.

The wood products that forests supply are so essential to society and the security of the state that several historians have attributed the rise and fall of many early Mediterranean and Middle Eastern cultures to changes in wood supply. Until use of coal became widespread in the 18th and 19th centuries, wood was the principal energy source—a title it retains in most of the developing world. Wood heat was required for cooking (to make grains digestible), home heating, and manufacturing the essentials of material culture. Pottery, bricks, glass, metal, and minerals like salt would not have been available in quantity without abundant fuelwood.

Until the mid-1800s all vehicles and ships—the basis for trade and exploration—were built of wood. Direct uses of the forest included pannage (the grazing of hogs), hunting, cattle grazing, and gathering honey. Given the importance of the forest, the repeated deforestation of so many regions seems shortsighted. Yet the cycle of deforestation, cultural decline, and reforestation was nearly universal from about 2000 B.C. until about 1800 A.D.

Babylon and Crete. As early as 2000 B.C. Babylon suffered wood shortages. Rented houses had no doors, and tenants had to take doors with them when they moved. Cedar was so rare that it was reserved for palaces and temples. Lack of wood forced King Hammurabi to seek additional sources, which were found in Crete. In Knossos large kilns fired pottery and smelters manufactured bronze from copper and tin for export throughout the region. When fuel became scarce in Crete, the Minoans imported wood from Mycenea, in what is now southern

The ancient forests of the Mediterranean were eliminated by a combination of human use and climate change. In the early 20th century many local communities in the region began to replant their forests. PHOTO COURTESY OF FAO.

Greece. As that forest was reduced and timber for shipbuilding became scarce, the golden age of Knossos and the Minoans ended.

Greece. The pollen record shows that pine forests once dominated Mycenea as far south as Pylos. By 1300 B.C. the population was expanding rapidly and the forest was being cleared for agricultural production. The government encouraged land clearing by forgiving the taxes of pioneer settlers. As the region's vegetation rapidly changed from pine forest to pasture for sheep, bronze workers and potters were forced to move north to find fuelwood. Deforestation and overgrazing in this semiarid region led eventually to flooding and soil erosion, and soil fertility declined. A similar fate befell the region near Troy, and the forests of Cyprus, too, were reduced to smelt copper. By 1200 B.C. much of the Mycenean world was experiencing depopulation. The cultivation of olives gradually replaced grain production, and the remaining population reverted to a subsistence economy. In the *Critias* (c. 350 B.C.) Plato described the result:

> What now remains compared with what then existed is like the skeleton of a sick man, all the fat and soft earth having wasted away and only the bare framework of the land being left.

Rome. The Romans deforested most of the central and southern Italian peninsula and much of Iberia (Spain) and North Africa. Their appetite for wood was enormous. The aqueducts were built with lime-based concrete and fired clay, both requiring substantial fuel. Grain production gradually exhausted the nearby soils and shifted to North Africa, reducing the native forests there. Much of Iberia was deforested for silver production. Wood shortages eventually encouraged glass recycling and the planting of willows for fuel and stakes for viticulture. But as scarcities increased, energy-intensive industries were relocated to southern Gaul and eventually Britain. The forests that were cleared to fuel the Roman ironworks regenerated during the early Middle Ages, only to be cut again: The 1085 survey known as the Domesday Book revealed extensive deforestation in southern England.

Muslims and Christians. The Muslims in Egypt traded with Venice for wood beginning around 1000, but this trade was eventually forbidden by the Pope. In *A Forest Journey*, Perlin reports that traders convicted of commerce with the "infidels" were threatened with excommunication. As the Venetians and the Ottomans vied for control of the Mediterranean, timber for ships was a matter of national defense, and the power of the trading city-states rose and fell as they won forested regions and then deforested them. In many cases, deforestation led to severe soil erosion and siltation. Several formerly important Mediterranean port cities are now located well inland because silt filled their estuaries. The port of ancient Antioch, for example, is now under many feet of waterborne silt.

A note about climate change. The deforestation of the Mediterranean basin and the Middle East is undeniably the result of human activities, but the landscape has also been altered by climate. Small changes in global temperature and moisture following the glacial advances of the early Neolithic period affected the viability of these early societies. A relatively wet period from about 5000 B.C. to just before 2000 B.C. ended with severe drought and desiccation. Climate change also produced wetter periods that favored food production for the Romans and Byzantines. The changing fortunes of several ancient peoples may therefore have been caused or amplified by climatic conditions.

Mesoamerica. The city-states of Central America provide an example of agricultural expansion and deforestation in the New World. The earliest Mayan settlements date from 2500 B.C. Although some domesticated plants are known from as early as 7000 B.C., improved maize was not developed until about 2000 B.C.

As population growth increased, cultivation gradually changed from swidden (slash and burn followed by long periods of fallow) to intensive forest clearing and continuous cultivation of hillside terraces and raised fields in wet lowlands. The most fertile soils were also the most susceptible to erosion. Because domesticated animals were rare in New World cultures, manuring was unknown. As soil fertility decreased and erosion increased, food production declined. Population probably peaked between 600 and 800 A.D. The exact causes of the Mayan collapse are unknown. Recently, some scholars have speculated that the changes were the result of atmospheric changes related to volcanic eruptions. Some environmental historians blame the ecological decline on the cycle of deforestation and agricultural erosion.

India. The civilization of the Indus Valley developed from the agricultural traditions of southwestern Asia. In many respects, its rise and fall parallel the experience of Sumeria. Cultivation of wheat and barley on irrigated fields supported the complex hierarchical society that flourished from about 2300 to 1900 B.C. The people constructed their buildings and monuments from oven-fired bricks, and wood fueled the ovens. In the warm climate, however, use of irrigation led to waterlogging and increased salinization. In a little less than 500 years, the Indus Valley civilization collapsed through a combination of salinization of the soils and deforestation.

China. Agriculture was practiced as early as 6000 B.C., with the domestication of millet and a small village system based on swidden. Urban, stratified societies emerged about 1750 B.C. Many of the soils in the northern region, where millet production was concentrated, were formed from wind-borne loess, which is highly erodible. The persistent flooding and siltation of the Yellow River, caused by forest clearing in the highlands, led to the decline of agriculture.

The Rise of Conservation

Evidence of early concern about natural resources comes from first-century B.C. Chinese records, telling of ancient rulers who sought to conserve soil, water, and wildlife as early as 1766 B.C. We find records of resource depletion in many ancient cultures, but real conservation of natural resources does not appear until the European Middle Ages.

Tension between forest clearing for food production and forest conservation emerged in the ninth century, at the time of Charlemagne. The Carolingians recognized the interrelationship between the growing populations, uses of the forest, and agricultural productivity. Wood was essential to the culture, and the wealth of the forest was distributed by a complicated system of property rights and duties held by combinations of towns, peasants, nobles, and the Church. Traditional rights of use for honey, beeswax, nut gathering, hog pannage, fuelwood gathering, toolmaking, resin collecting, and barrel making were vested in many communities, and royalty held rights to hunt. It was the balancing of royal rights with the rights of the community and the demands of the Church that forced the evolution of forest customs and laws.

From 1100 to 1200 A.D., laws restricting the uses of the forest had become common throughout England, France, and Germany. Protection of reserved forests for royal hunting became more common,

Some of the earliest regulations of land use practices in Europe date from the 1100s, when villages began to regulate grazing in forests and on meadows in an effort to prevent erosion and flooding. Livestock grazing and cultivation on steep hillsides in Andermatt, Switzerland, have been controlled for hundreds of years. PHOTO COURTESY OF ROBIN MAILLE, USDA FOREST SERVICE.

but there are also examples of Church or state orders for protection of forests to provide fuel for heating, cooking, and industry, especially salt production. It is about this time that foresters appear as the officials responsible for enforcing the decrees.

Concern for protection of the watersheds of alpine valleys also emerges in the Middle Ages. Control of livestock grazing (particularly goats) and cultivation in the steep terrain of France, Switzerland, and southern Germany predates 1400. There is evidence of control of livestock grazing in German forests as early as the 1100s.

But the forests of Western Europe recovered without help from early regulators. Just as forest clearing reached its historic maximum c. 1300, famine and disease spread through the region. The limited kinds of crops were vulnerable to weather, insects, and plant disease. Severe famines from 1315 to 1317 were followed by outbreaks of human disease. The Black Death further reduced European populations. As human populations declined, in many areas the cleared land reverted to forest.

True, purposeful conservation—the wise use of natural resources to meet the needs of the *current* generation—did not find effective implementation until scientific knowledge about natural processes combined with technology made it possible to limit some of the adverse effects of human-caused environmental change. The 200 years from 1600 to 1800 witnessed the first systematic treatment of the ideas associated with conservation of natural resources in Western literature.

In 1662 John Evelyn's *Sylva* was commissioned by the Royal Society of Britain. The first modern forest policy report, it analyzes the causes of deforestation and explores the potential for reforestation. Evelyn attributed deforestation to agricultural expansion and industrialization. Iron and glass manufacture, demand for home heating, and forest clearing for agricultural production combined to limit the availability—and increase the cost—of wood fuel. Defense policy was also an issue because England had lately been forced to seek ship timbers from the Baltic states. Political instability in England itself compounded the wood supply problem.

Evelyn proposed a rational land-use policy that championed efficient allocation of resources to meet social needs. His ideas prefigured the utilitarian philosophies associated with late-19th-century American conservation leaders like Fernow, Powell, McGee, and Pinchot.

Timber famine and national commerce policy also motivated Jean Baptiste Colbert, a statesman who served King Louis XIV during the 1660s. In the Ordinance of 1669, which followed an eight-year study of French forest policy by 21 commissioners, he explicitly acknowledged the importance of meeting the needs of future generations. The ordinance was comprehensive and treated the harvesting of trees, reseeding requirements, livestock grazing, and the manufacture of ash and charcoal.

North American Conservation

When European colonists arrived in the 1500s, they regarded the American forests as primeval, but in fact the land had already been shaped by several thousand years of human habitation. Native Americans had changed the forest to favor certain plants and wildlife and cyclically cleared it for agriculture. Clearing and burning were common practices. Recent estimates suggest that perhaps 20 million acres of eastern North American forest had been converted for food production by the time of European contact.

Our knowledge of the effects of Native Americans on ecology remains imperfect, however. Population densities are only roughly estimated, as are the proportions of fish, game, and agricultural products in the diet. There is also much speculation about the coincidence of the extinction of large Ice Age mammals with the increase in human population during the late Pleistocene, as well as debate over the archaeological evidence of deforestation and salinization of soils in the arid Southwest.

There is little doubt, however, that Native Americans altered their environment and depleted natural resources, at least at the local scale. On a continental scale, they lacked the population density and the technologies to make the kinds of changes that would come with European settlement. And because of climatic differences, the early deforestation of North America is not as apparent as that of the arid Mediterranean, Middle East, or China. In much of New England and New York, in fact, forests now grow where no forest existed 100 years ago.

Protection of the spiritual aspects of nature was important for Native Americans. But for the new arrivals, especially those who came in the first waves of immigration seeking religious freedom, the forest was an obstacle to security, commerce, and spiritual salvation—attitudes that were the legacy of late medieval Christian values. The forest was the home of the ancient gods, of the old morality. It was the duty of Christian men to clear the forest and convert it to a garden. Particularly for the fundamentalist Protestants who dominated New England, converting the forest would bring salvation and economic prosperity.

The forest harbored perceived dangers—Native Americans, wild animals, the unknown—and its quick conversion to fields, pastures, and woodlots provided security through increased food production and elimination of predators. Because the forests were necessary for fuel, pannage, and pasture, however, the colonists did not clear them altogether. Instead, they sought a balance. Where soils were productive, agriculture was established. In areas with less fertile soil, lumbering and wood-fueled industry (potash and iron smelting) offered economic rewards. Thus the European settlers reduced the forest area and converted what remained to more readily available economic assets.

By 1800, the pursuit of material well-being was being challenged by a rival belief that nature was "morally instructive." This idea, different from the rational conservation of Evelyn and Colbert, is critical to understanding the modern concept of sustainability. At the same time that clearing and exploiting the forest were making North Americans prosperous, the Romantics—artists, writers, and poets—were praising the unspoiled virtue of people who learned their life skills from natural processes rather than human civilization.

The political power of the idea that nature is morally instructive was already apparent in the writings of William Penn and Thomas Jefferson. Penn sought to save the forested resource in his new colony by prescribing that settlers retain one acre in forest for every five acres cleared. Jefferson considered the yeoman farmer an ideal citizen and favored a government in which local citizens came together, face-to-face, to resolve common issues. (The concept of direct participation by "stakeholders" emerges again 200 years later as a cornerstone of social sustainability.) Inherent in his ideal of governance was a distrust of commercial and industrial power—a theme echoed today in environmentalists' concerns about globalization.

Now, in the 19th century, James Fenimore Cooper and Washington Irving were writing novels that celebrated the primitive and the wild. Transcendentalists like William Cullen Bryant, Ralph Waldo Emerson, and Henry David Thoreau held that all men through their inner perceptions were one with each other and one with nature, and that nature was one with God. Because nature was now linked with spirituality, nature must be preserved to allow men a closer link to their religion.

The rediscovery and celebration of the forest as sacred was a reaction to 18th- and 19th-century urban conditions and industrialization. Rapid population growth, clearing for agriculture, and new technologies had led to rapid deforestation in much of Europe. In North America, cities were beset by problems of disease caused by water pollution and poor sanitation. In the East the forests had been cleared and fuelwood prices were climbing. Limited federal reserves for strategic forest products like ship timbers, pitch, and turpentine had been public policy since the early 19th century, but most federal land had been given away or sold to promote settlement and material progress. In short, urbanization presented physical, political, and social challenges, and technology seemed slow to meet them.

Linking popular support for nature preservation with the abstract philosophy of the Transcendentalists was John Muir. As an outspoken

By the early 19th century, many artists were producing landscapes that emphasized the grandeur of nature and the nobility of native peoples. Paintings of the Hudson River School, like "Falls of the Kaaterskill" (1826) by Thomas Cole, reflect a shift in the relationship between humans and nature. PHOTO COURTESY OF WARNER COLLECTION OF THE GULF STATES PAPER CORPORATION, TUSCALOOSA.

advocate of nature preservation, Muir galvanized public support for the protection of wild forests as a place for spiritual renewal. Some public lands had been set aside for health benefits in the 1840s, notably at the Hot Springs in Arkansas. Now, as the preservation movement gained strength, the spectacular scenery of Yosemite Valley was secured by a federal grant to the state of California in 1864. Creation of Yellowstone National Park in 1872 and the Adirondack Preserve in New York State in 1885 soon followed. In western Canada, creation of the Hot Springs Reserve in Banff in 1885 marked the beginning of the Canadian National Parks system.

The morally instructive view of nature had evolved from a literary and artistic movement to public policy, but it would face its greatest challenge from two other gospels—efficiency and progress.

Utilitarian Conservation

The utilitarian approach to conservation gained political support in the United States after the Civil War. Its proponents argued that the nation could meet the needs of the current generation with efficient scientific management by technically trained professionals who would make decisions for the public. Utilitarian conservation was associated with the Progressive movement, which sought to meet the challenges of industrial capitalism by expanding governmental influence.

During the second half of the 19th century, many people believed that plowing the native prairie and planting trees on the Great Plains would change the climate and increase rainfall. The government gave away land to settlers who would plant and maintain forests on the arid plains. PHOTO COURTESY OF NEBRASKA STATE HISTORICAL SOCIETY.

The almost religious belief in the power of science to solve problems was kin to manifest destiny—the conviction that God intended the United States to occupy North America from the Atlantic to the Pacific. Canada had its manifest destiny, too: The government encouraged settlement by "loyal British subjects" on the western frontier. In *Americans and Their Forests*, Michael Williams suggests that "To fell the forest was almost to enter the kingdom of heaven on earth, as the making of new land seemed to demonstrate the direct causal relationship between moral effort, sobriety, frugality and industry and material reward." Converting the former wilderness to a landscape of fruitful farms and pleasant villages became a national priority.

New technologies allowed the conversion of the Great Plains to farms. The railroads recruited emigrants in Europe and sold them dry prairie land. Enthusiastic boosters proclaimed that "rain followed the plow," and according to Stegner (1953), the idea that settlement, plowing, and tree planting would modify the climate, evaporate more water into the air, and magically milk it from the clouds as rain flourished, until the drought of the late 1880s killed it. Such ideas reflected the tremendous (and sometimes misplaced) faith in the ability of technology and science to conquer nature and secure its benefits for mankind. Throughout the 20th century, people on the Great Plains continued to plant trees for windbreaks and soil conservation.

As faith in progress and technology grew, Americans also became aware that their forests were disappearing. Many had read George Perkins Marsh's *Man and Nature*, published in 1864.

A flurry of official reports and popular pamphlets on the deforestation of the United States and Canada followed. Increase Lapham documented the deforestation of Wisconsin. Frederick Starr addressed the economic consequences of deforestation. In Canada, lumberman James Little published a pamphlet in 1872 calling for forest conservation. Finally, in 1873, Franklin Hough read a paper at the American Association for the Advancement of Science that catalyzed the leading scientists of the day to ask Congress to create a federal commission on forests.

Congress responded three years later, creating a Division of Forestry within the Department of Agriculture. Hough became its head and authored widely read reports on forestry. Practical programs for forest

The publication in 1864 of George Perkins Marsh's *Man and Nature, or Physical Geography as Modified by Human Action,* led to new interest in how humans affect the environment. A U.S. congressional representative and diplomat, he believed that humans had an obligation to restore the land through resource management. PHOTO COURTESY OF DAGUERREOTYPE COLLECTION, LIBRARY OF CONGRESS.

Franklin Hough helped lead the early U.S. effort for forest conservation and was the first federal forestry agent. PHOTO COURTESY OF USDA FOREST SERVICE.

German-born Bernhard E. Fernow helped establish early forestry colleges in the United States and Canada and did much to shape the initial direction of professional forestry in North America. His contributions have been eclipsed by those of Gifford Pinchot, who had superb managerial skills and wielded more political influence. PHOTO COURTESY OF FOREST HISTORY SOCIETY COLLECTION.

Following the Civil War, the cause of forest conservation gained popularity in North America. The First Forest Congress was held in Cincinnati in 1882. Thousands attended the event, which included ceremonial tree plantings. The Forest Congress was sponsored by the American Forestry Association (now known as American Forests). SKETCH BY J. KNAPP, *FRANK LESLIE'S ILLUSTRATED NEWSPAPER*.

protection were introduced. In 1875 several prominent Canadians helped establish the American Forestry Association (now American Forests and the oldest citizen conservation group in North America). The first American Forest Congress was held in Cincinnati in April 1882. The U.S. federal government created forest reserves in 1891 to improve and protect the forest, protect watersheds, and supply timber to local settlers. Canada identified forest reserves in 1894 and set up a system of Dominion Lands beginning in 1911.

But a philosophical split was already apparent. Some conservationists favored the complete protection of public forests in reserves where timber harvest would be forbidden. Others, the utilitarians, suggested that the application of science and technology could lead to efficient management and use of the forest to meet society's economic and material needs.

Bernhard E. Fernow, a German forester, was among the latter. In 1886, he became director of the Agriculture Department's Division of Forestry. Fernow was generally opposed to federal regulation of private forests and chose to cooperate with timber companies to promote efficient use and management. He also promoted research on the uses of wood products and the development of professional forestry.

Fernow's influence on conservation policy in the United States and Canada was profound. He transplanted the prevalent European ideal of forest management, which was rooted in agriculture. It specified efficient, perpetual management of the forest to meet national needs for "crops" of wood and water. Other uses might be allowed to the extent that they did not interfere with timber production and watershed protection. Fernow eventually became the secretary of the American Forestry Association, established the nation's first (and short-lived) university-based school of forestry at Cornell University, and led the forestry program at the University of Toronto.

Fernow's successor was a politician and forester who hitched the future of America's forests to a growing political movement. Gifford Pinchot was by all accounts a brilliant zealot. The scion of a wealthy mercantile family, he blazed a bright and often controversial path for American forestry at the turn of the 20th century.

Perhaps more than any other individual, Pinchot is associated with the idea of efficient forest management to meet the needs of the people. Echoing English jurist and philosopher Jeremy Bentham (d. 1832), he called for the greatest good for greatest number over the long term and used his considerable talents to influence the media, Congress, and the president on forest policy and management. He favored federal regulation of private forests and advocated managing public forests so that they would become self-financing through the sale of timber.

Pinchot was closely associated with the Progressives. Their movement, which lasted from about 1870 to 1920, was a response to the growing wealth and political power of industrial capitalism. The Progressives sought to counterbalance industrial economic power—the large corporations and trusts that controlled the railroads, steel, and oil and were considered a threat to democracy—with an expanded, activist government. This government would be administered by an educated managerial class immune from the influences of the trusts and big-city political bosses. Among the Progressives' legacies are the federal antitrust laws, public health programs, the civil service, direct election of U.S. senators, and the use of petition to place issues on the ballot.

The Progressives shaped U.S. conservation policy in the late 19th and early 20th centuries through centralized planning for natural resources, retention of public lands, decision making by professional managers, and self-financing for many conservation programs. Most of our modern state and federal natural resource management agencies were created during this era, and their programs and management philosophies are the direct result of Progressive thinking.

Public forest managers sought the efficient, perpetual production of forest products. Out of fear of a timber "famine," the federal forest reserves were gradually transformed into national forests. At first the emphasis was on fire control and reducing overgrazing, but Pinchot's goal was to demonstrate that "forestry pays." With that in mind, the Forest Service set out to demonstrate the viability of commercial timber production from public lands.

That forests could be used for a variety of purposes—timber, fuel, hunting, pasturing, and so-called minor forest products like nuts and mushrooms—was well known, and 20th-century "multiple-use" forest management arose from a centuries-old tradition. The utilitarian conservationists adopted the doctrine of multiple use and the idea of sustained yield. They believed regulation and scientific management of the forest would perpetually produce products for the benefit of all citizens.

Because timber was the most valuable of those many uses, most of the effort was placed on managing the forest for wood products. By about the end of World War I, timber production on the national forests returned more revenue to the Treasury than livestock grazing. Wood supply to meet economic priorities became an increasingly important function of the public forests.

The supremacy of timber and faith in sustained-yield timber man-agement in Canada, too, are apparent in a 1910 statement of forest policy:

Our legislators are well aware that forests feed springs, prevent floods, hinder erosion, shelter from storms, give health and recreation, protect game and fish and give the country aesthetic features. However, the Dominion Forest Reserve policy has for its motto: 'Seek ye first the production of wood and its right use—and all these other things will be added unto it.'

Ecological Science and Scientific Forestry

Beginning in the 1910s and 1920s, some forest reserves were redesignated as national parks. Anxious about these transfers to the National Park Service, the Forest Service adopted multiple use as a catchword to convince Congress and the American people that it, too, could provide recreational opportunities—fishing, hunting, wilderness, and camping. Promotion of multiple use increased dramatically following World War II, but so did the demand for lumber for housing. By the 1950s and 1960s the effects of increased timber harvests were readily apparent to the many Americans who came to their national forests for outdoor recreation, and they complained to their legislators.

In 1960 the Multiple Use and Sustained Yield Act legitimized what had been agency policy for years. The forests would now be managed for outdoor recreation, water, and wildlife habitat, as well as for timber and livestock grazing. Four years later, after a struggle of more than 20 years, Congress passed the Wilderness Act. The national forests were now a preserve, timber supply, wildlife refuge, public pasture, and picnic ground rolled into one. With the legacy of Fernow's and Pinchot's faith in modern scientific management, it was the foresters' job to provide all those products and services.

At the same time, expanding knowledge of ecological systems and processes began to influence public forest management. The environmental movement was challenging the way foresters managed public forests, and the ideological struggle between utilitarian conservation and preservation of nature for moral instruction—renamed spiritual renewal—intensified. Increasingly the environmental community saw the science of ecology as a way to change the focus of public forest management from product management to protecting ecological processes.

Ecology was a relatively new discipline. Description (or natural history) had been the basis of the natural sciences until the 1840s and 1850s, when Charles Darwin and Alfred Russell theorized that the development of species was a process shaped by natural selection. The idea that life on earth was a dynamic process—and that geological processes were continuing to shape the earth itself—stirred a storm of controversy about the relationship between man and God and the literal interpretation of Christian and Hebrew texts. In the 1860s German zoologist Ernst Haeckel coined the word "œcology" to describe the study of the relationships among plants and animals and their physical environment.

By the early 20th century, the science of ecology was fully rooted in North America and Europe. The concept of an ecosystem was developed in the 1930s by English ecologist George Tansley. Following World War II, the science of ecology bloomed. Driven in part by interest in how nuclear radiation moved through the environment, the federal government increased funding for basic ecological research at Oak Ridge, Tennessee, and in university biology departments. Knowledge of how nutrients and energy moved through the environment expanded rapidly and changed the way scientists (and eventually nonscientists) thought about the earth. Research based on descriptive natural history was replaced by research in ecosystems and processes—and the new research showed that some species had become extinct, others were threatened and still others were increasing as a result of habitat alteration.

As ecological science developed, the negative effects of some kinds of intensive timber management became more apparent. The economically efficient method of clearcutting in the old-growth forests of the Pacific Northwest, the northern Rocky Mountains, and West Virginia polarized the public. Despite the fact that clearcutting is sometimes the best method for regenerating some tree species, it remains controversial and has been significantly reduced on the national forests and other federal public lands. The federal forest agencies began to change their management paradigm. Multiple use gave way to "new perspectives" and then to "ecosystem management." The change was slow, in part because of the tremendous economic consequences to the communities that depended on forest products and livestock forage.

Our growing understanding of ecology and life as a process meant that eventually our approach to forest management would have to change as well. The systems approach that advanced our knowledge of biological science would also begin to influence our ideas about the economy. Gradually these ideas would come together to form the basis for modern sustainability.

MAKING THE CONCEPT
A REALITY

The modern search for sustainability began more than 200 years ago with the same two issues that drive the modern debate—human population and the ability of the earth to produce the natural resources on which we depend. In the intervening years our knowledge of biology, ecology, and meteorology has grown rapidly, but our ability to resolve the problems related to human behavior has not kept pace. Ultimately, sustaining the forest is as much about controlling human behavior as it is about ecological science—perhaps more.

Recognizing Limits

At the end of the 18th century, historians were curious about the population of the earth in ancient times and what had happened since. The questions were not new. The ancient Greeks believed that as the earth aged, it had become less productive and supported fewer people in a poorer state than during a supposed "golden age." Now, intellectuals were suggesting that human population of the late 18th century was less than it had been in ancient times because of wicked tendencies in Western civilization. Well-known figures such as Hume, Montesquieu, and Sir Walter Raleigh published essays on the subject.

One of the principals in this debate was Robert Wallace. In 1761 he published *The Various Prospects of Mankind, Nature and Providence,* arguing that in a utopian state, human population would increase so rapidly that it would outstrip its food supply. Wallace's argument inspired Thomas Malthus.

Malthus published his *Essay on Population* in 1798. It remains one of the most important and controversial works in the English language. Malthus derived his argument from two principles: first, that food is necessary, and second, that passion between the sexes is a permanent human trait. The population of all living things therefore tends to increase geometrically as long as there is sufficient food. But food production is limited by land capability, and when the land is fully occupied by agriculture, it will no longer be able to produce enough food to feed the growing population.

Malthus's work popularized the idea of environmental limits. It was controversial for the same reason, and Malthus was widely criticized by church leaders and utopians. In the period after its publication, the conclusions of the *Essay on Population* seemed wrong: There was an unprecedented growth in human population in Europe and the Americas as well as an unprecedented increase in the standard of living.

That Malthus's predictions did not come true was the result of two factors. The first was the "discovery" of the Americas by Europeans. The expansion of agriculture to previously uncultivated areas effectively increased global food production, and migration to the New World relieved some of the population pressure in Europe. The second and perhaps more enduring factor was the development of new technologies.

Modern resource economists suggest that in the past 400 to 500 years, the world has experienced a "boom" unprecedented in human history. Technologies of all kinds—energy use, biotechnology, medical discoveries—have boosted food production and increased human life spans. Thus the population has increased rapidly and we now find ourselves debating the same issues that Malthus and other economists confronted two centuries ago.

Spaceship Earth

The systems perspective that energized the physical and biological sciences began to influence economics during the second half of the 20th century and led to the notion of the earth as a closed material system. That is, Earth is much like a spaceship: It contains all the material resources available for sustaining all forms of life, and except for energy and knowledge, no other resources are available to us. In the 1950s and 1960s systems economists like Kenneth Boulding said that the best measure of human well-being was not how much stuff (resources) we used but the condition of the resources not yet used. Picking up this idea, many environmental and ecological economists began to argue that the goal of economies should change from maximizing current consumption to optimizing the efficiency of consumption in order to sustain the quality of life over the long term. Like Malthus's essay on population growth, systems theory and the idea of optimizing resources over time would shape our current conception of sustainability.

In 1972 there appeared a controversial report called *The Limits to Growth*. This report—written by Donella and Dennis Meadows and several others and commissioned by the Club of Rome, an informal group of academic, government, and business leaders—identified five factors that "determine, and therefore, ultimately limit, growth on this planet": population, agricultural production, natural resources, industrial production, and pollution. Using then-sophisticated computer programs, researchers modeled the earth's economy, resource use, population growth, and pollution as a system. Some of their assumptions about the limits of the physical world were similar to those of Malthus and also found an antecedent in John Stuart Mill, who in the 1800s had written about the possibility of a steady-state economy— one with little or no growth. The Club of Rome report was generally skeptical of the ability of technology to solve problems related to industrial pollution and growth but saw hope in a steady-state economy.

Also in 1972 the United Nations held a conference in Stockholm on the human environment, at which workers in international economic development discussed how some projects designed to alleviate poverty in the developing world were damaging both the physical environment and the cultures of local people. The concept of ecologically sustainable development, or "ecodevelopment," soon became current.

Those two ideas—a physically limited earth and the negative effects of large-scale international development projects—are the intellectual parents of the modern concept of sustainable development.

A variation on this theme emerged from other economists who suggested that economic growth was still the only hope for alleviating poverty in the developing world. In this view, economic growth that matched but did not exceed the productive capacity of the ecosystem would be "sustainable growth." The term is controversial because of its embedded assumptions. Even at a very slow rate of increase, no rate of growth can be sustained forever because of the physical limits of the earth as a natural system.

Global Concerns

The U.N. World Commission on Environment and Development published its report, *Our Common Future,* in 1987. The Bruntland Report, as it is more commonly known, popularized sustainable development as a term expressing the need to unify ecological, economic, and social development goals. The report suggests that sustainable development (that is, development that meets the needs of the present without compromising the ability of future generations to meet their own needs) is a necessary first step toward resolving the cycle of poverty and environmental degradation that grips many developing nations.

In a thoughtful critique, Bartlett (1997–98) asks whether it is indeed possible to increase economic activity without also increasing the rate of consumption of nonrenewable resources. Economic growth, even at a relatively low rate of increase, creates very large effects in a fairly short period of time. To be sustainable, economic development must be limited by the ability of renewable resources to grow and be replenished. Sustainable development is possible only if the rate of consumption is less than or the same as the replacement rate of the resources used to feed the economic growth. In nations whose population is increasing by 3 percent each year, renewable resource production and consumption must grow by the same amount just to maintain the current standard of living.

In 1983 Gro Harlem Bruntland established and chaired the World Commission on Environment and Development (known as the Bruntland Commission), which promoted the concept of sustainability. A physician by training, she served as prime minister of Norway three times in the 1980s and 1990s and became director-general of the World Health Organization in 1998. PHOTO COURTESY OF NORWEGIAN EMBASSY.

Most multinational conventions are negotiated through a consensus process. As a result, they contain very broad agreements that are often based on nebulous terms. The concept of sustainability has great appeal to diplomats who must negotiate international protocols, but it contains a logical flaw. Strictly speaking, sustainability is possible only in a system that relies completely on renewable resources. In some areas, demand for renewable resources is greater than the ability of those resources to be renewed, and nonrenewable resources are being used up. Sustaining resources while making some of them available for consumption is therefore a challenge for environmental managers.

When the U.N. Conference on Environment and Development (UNCED) met in Rio de Janeiro in June 1992, sustainability of the world's forests came into focus. The 178 nations at the UNCED conference adopted a set of "forest principles," and two treaties on climate change and biodiversity were signed by many heads of state.

The Forest Principles had originated in July 1990, when the heads of the major economic powers (Great Britain, France, Germany, Italy, Canada, Japan, and the United States, known as the G7) met in Houston. Then-President George Bush suggested that negotiations begin on a legally binding forest convention. But significant differences between developed and developing nations made such a convention impossible. Instead, the Forest Principles were adopted by consensus at the Rio meeting. These principles call for sustainable management of forests while recognizing national sovereignty:

> Forest resources and forest lands should be sustainably managed to meet the social, economic, ecological, cultural and spiritual human needs of present and future generations. These needs are for forest products and services, such as wood and wood products, water, food, fodder, medicine, fuel, shelter, employment, recreation, habitats for wildlife, landscape diversity and other forest products. Appropriate measures should be taken to protect forests against harmful effects of pollution, including air-borne pollution, fires, pests and diseases in order to maintain their full multiple value.

The essence of multiple-use management is apparent in this statement. But except in Europe, North America, India, Maylasia, Ghana, Japan, New Zealand, Australia, and a few other countries, the majority of natural forests are either inadequately managed or not managed at all. Many nations lack the political, legal, and property mechanisms and institutions—laws, agencies, markets, honest officials—necessary to protect and manage their forests.

Canada provides an interesting case of a developed country that has embraced sustainable forestry. Led by the Canadian Council of Forest Ministers, Canada has placed sustainability at the heart of its national forest policy, with broad public participation and coalition support. The policy establishes ten model forests in Canada (and several others in other nations) where the principles and mechanisms of sustainable forestry can be demonstrated. The federal and provincial governments are shifting their forest research agendas to deal with problems of sustainability. An important part of the Canadian strategy is periodic review of the program by independent experts. In the early 1990s Canada became a leader in the effort to assess progress toward the goals of sustainable forestry through criteria and indicators.

Criteria and Indicators

How will we know the sustainable forest when we see it? Sustainable forestry relies on a system of criteria and indicators. Criteria are the goals; indicators are measurable signs that the goals are being achieved.

The use of criteria and indicators to assess sustainability in forestry goes back to 1990, when the International Tropical Timber Organization (ITTO), which had originally been established to organize international trade in tropical timber, issued guidelines for the sustainable management of natural tropical forests. These were followed by guidelines for management of planted tropical forests and conservation of biodiversity in tropical forests managed for wood products.

After its Conference on Environment and Development in 1992, the United Nations formed a commission on sustainable development to oversee a panel on forests, called the Intergovernmental Forum on Forestry, which met several times to discuss sustainable forestry and the possibility of a binding global forest convention. To date, there is no consensus on pursuing a binding international protocol on sustainable forest management. However, these international efforts have promoted criteria and indicators as a way to assess progress toward sustainable forestry.

At least seven groups of regional governments, in efforts generally called processes, have now formulated criteria and indicators of sustainable forestry. Each criterion identifies a specific element of forest sustainability, and each indicator is a specific way to measure one of the criteria. The regional process approaches include four to seven criteria and 25 to 70 indicators. Because defining and measuring progress toward sustainability require change at the local, national, and international levels, some of the regional agreements address all three spatial scales.

Six of the regional approaches are as follows:

The Helsinki Process (the Pan-European Process on Criteria and Indicators for Sustainable Forest Management) covers forests in the boreal, temperate, and Mediterranean zones. It is convened by the Ministerial Conference on the Protection of Forests in Europe. In 1998 it adopted six national-level criteria.

The Montreal Process on Criteria and Indicators for the Conservation and Sustainable Management of Temperate and Boreal Forests addresses the United States and Canada and 10 other countries. This conference adopted seven criteria and 67 indicators, set down in the Santiago Declaration of February 1995. Its decisions are not legally binding.

The Tarapoto Proposal for Criteria and Indicators for Sustainability of the Amazon Forest was adopted in February 1995 in Tarapoto, Peru. This conference proposed three sets of criteria and indicators. At the global level there are one criterion and seven indicators. At the national level there are seven criteria and 47 indicators. At the forest management unit level there are four criteria and 22 indicators.

The Dry-Zone Africa Process originated in a meeting in Nairobi in 1995. This conference identified seven national-level criteria and 47 indicators of sustainable forestry. Subsequent meetings have been held to endorse the proposals within the region.

The Near East Process began in Cairo in October 1996. Participants developed seven national-level criteria and 65 indicators for sustainable forest management; they have since met in Cairo and Damascus to review progress and promote testing and implementation.

The Lepaterique Process of Central America was initiated following a meeting in Tegucigalpa, Honduras, in January 1997. The seven Central American nations identified eight criteria and 52 indicators for app-

Summary of Sustainable Forestry Criteria in International Agreements

Criteria and indicators	Helsinki	Montreal	ITTO	Tarapoto	Dry-Zone Africa	Near East
Levels						
Forest unit level	no	no	yes	yes	no	no
National level	yes	yes	yes	yes	yes	yes
Global level	no	no	no	yes	no	no
Forest resources						
Extent of forests	yes	[1]	yes	[2]	yes	yes
Global carbon cycles	yes	yes	no	no	[3]	no
Ecosystem health and vitality	yes	yes	no	—	yes	yes
Biological diversity	yes	yes	[4]	yes	yes	yes
Forest functions						
Productive functions	yes	yes	yes	yes	yes	yes
Protective and environmental functions	yes	yes	yes	yes	yes	yes
Development and social needs						
Socioeconomic functions and conditions	yes	yes	yes	yes	yes	yes
Institutional framework						
Policy and legal framework; ability to implement sustainable management	yes	yes	yes	yes	yes	yes

Table 1

[1] In the Montreal Process, Forest Resource is considered not a criterion but an indicator for other criteria (conservation of biological diversity and maintenance of the productive capacity of forest ecosystems).
[2] In the Tarapoto Proposal, the criteria Extent of Forest Resource and Biological Diversity are combined into one criterion, Conservation of Forest Cover and Biological Diversity.
[3] In the Dry Zone Africa proposal, the criteria Global Carbon Cycles and Extent of Forest Resources are combined into one criterion.
[4] The ITTO developed supplementary guidelines for biodiversity rather than include it as a criterion.
SOURCE: ADAPTED FROM FAO 1997.

lication at the national level and four criteria and 40 indicators for application at the regional level.

Table 1 compares the criteria developed by six processes and reveals significant agreement. Almost all the regional approaches address institutional and social sustainability as well as the productive and protective functions of the forest. All address important biophysical issues, including conservation of biological diversity, global carbon cycles, ecosystem health, and the extent of the forest resource. Several governments and nongovernmental research organizations are now testing the criteria and indicators approach at the field and national levels.

Such criteria and indicators are used to try to define sustainable forestry. They represent a broad consensus about the elements of the concept and provide a framework that governments, industries, and interest groups can use to assess progress toward the goal. Regional consensus on criteria and indicators is an important first step toward forest sustainability. But there is much work that remains to be done.

Forest Certification

In addition to intergovernmental efforts to establish criteria and indicators, a number of industry organizations and nongovernmental organizations have developed programs to certify that forests are well managed. Private landowners, state and local governments, and forest industries are participating in these programs. Certification is a general term used to describe the process of verifying that forests are well-managed and communicating that through some type of recognition.

Certification is an attempt to use market forces rather than governmental regulation to achieve sustainability. The original assumption was that educated and concerned consumers would pay a premium for wood products that came from well-managed forests. More recently, proponents of certification have deemphasized the idea that the consumer will pay a premium and have concentrated on market access or market share as justification.

In the United States, Canada, and much of Europe, managers of public and private forests must often comply with strict environmental regulations. But skepticism about the effectiveness of governmental regulation has given not-for-profit organizations and industry trade organizations opportunity to promote sustainable forest management beyond legal requirements. Many forest products companies and other forest landowners seek to reassure their customers that their products are the result of good forest stewardship. But because forest sustainability is difficult to define, there has been little agreement on precisely what constitutes a sustainably managed forest. Initially, several groups attempted to certify "sustainability," then changed their approach to certify that forests were "well managed."

Like the auditing process that an accountant uses in reviewing financial performance, certification assures stockholders and stakeholders that managers are meeting appropriate standards. The standards may derive from international criteria and indicators of forest sustainability. The assurance may come from the company itself (first-party certification), from a trade or landowner association (second-party), or from an independent body that provides certification (third-party). In the United States, Canada, and Europe all three approaches are used.

Several organizations have developed forest certification standards. The Forest Stewardship Council (FSC) is an international, non-governmental organization set up to evaluate, accredit, and monitor groups that offer forest certification. It accredits these organizations to guarantee the authenticity of their assessments. Headquartered in Mexico, FSC has been instrumental in accrediting for-profit and not-for-profit organizations engaged in forest certification, and in doing so, it attempts to promote a worldwide standard for principles of forest management. A variety of private forests and public forests have now been certified by FSC-accredited organizations, including tribal forests in Wisconsin, state forests in New York and Pennsylvania, local public forests in Minnesota and Massachusetts, and private forests in many states. The majority of FSC-certified lands, however, are outside North America and mostly plantation forests.

Because the current certification processes are expensive, most of the forests that have been certified are large areas owned by governments and industries. Many owners of smaller forests remain interested in forest certification, but the cost compared with the potential benefits remains a question. Organizations such as the American Tree Farm System and the National Forestry Association (the Green Tag program)

A variety of nongovernmental organizations offer certification for sustainably managed private and public forests. Here, the President of the National Forestry Association presents its "Green Tag" for a management unit on the Pisgah National Forest. PHOTO COURTESY OF STEVEN ANDERSON.

Foresters at the Seven Islands Land Co., in Maine, have made certain that the next generation of trees is well established before they harvest the older trees that make up the overstory. Seven Islands is certified under both the Sustainable Forestry Initiative (SFI) program with AF&PA and the Forest Stewardship Council (FSC) guidelines. PHOTO COURTESY OF SEVEN ISLANDS LAND COMPANY.

offer certification programs for nonindustrial forest owners, including those with smaller acreages. Recently, the National Forestry Association certified a portion of a national forest in North Carolina

In the United States, one of the largest certification programs is sponsored by a trade association. The American Forest and Paper Association (AF&PA) established the Sustainable Forestry Initiative (SFI) in 1995 and made adherence to its standards a requirement for membership. As a result, several forest products firms chose to leave the association, but the remaining AF&PA members own 90 percent of the industrial forestland in the United States and all have agreed to implement the program's standards. Using a system much like the criteria and indicators that characterize international agreements, SFI relies on a system of principles, objectives, and performance measures. It incorporates a review by an expert panel, and several member companies are now seeking independent, third-party field audits of their management practices.

There are many examples of sustainable forestry programs; the three presented below help illustrate the process of translating sustainable forestry from a concept to on-the-ground management.

Seven Islands Land Company manages more than 900,000 acres for the Pingree family, which has owned land in Maine for more than 150 years and is committed to the long-term conservation of its forest resources. In 1993, on Seven Islands' recommendation, the Pingrees sought and gained third-party verification. At the time, it was the largest certified forest in the United States. The certifying company, Scientific Certification Systems, commented that Seven Islands demonstrated the highest principles of sustainable forestry, ecosystem stewardship, and social responsibility and praised the company for sound forestry, financial stability, and reliability.

Seven Islands emphasizes biodiversity in the forests it manages. Wildlife management has been carefully integrated into forest management with particular concern for cavity-nesting birds, deer yards, riparian areas, and protection of rare and endangered species. Natural regeneration of native species is a guiding management principle. Seven Islands leaves standing high-quality trees that will enhance the growing stock. Market demand does not have strong influence on the harvest rate of the Pingree land.

Seven Islands takes sustainable management beyond the forest by monitoring the production processes used to manufacture flooring and shingles. The sawmills and lumberyards that it supplies are becoming certified as well.

McGregor Model Forest provides an example of sustainable forestry in central British Columbia. Like much of the forestland in the Canadian west, the McGregor Forest is public land, owned by the province and managed under a long-term lease to Northwood, a forest products company. The 181,000 acres of woodlands produces an average of 386,000 cubic meters of timber annually, which is harvested year-round by both partial harvesting and clearcutting practices. Northwood shares inventory information and takes management guidance from the McGregor Model Forest Association, a collaboration of 33 interested organizations.

The McGregor approach to sustainable forestry identifies biodiversity, forest productivity volume, forest health, conservation of water and soil, carbon storage, and community socioeconomic gain as criteria. Indicators are used to monitor the success or failure of management practices to meet those criteria and improve future management. This is known as adaptive management, and it recognizes our imperfect knowledge and the need for flexible management strategies.

Westvaco Corporation owns 1.238 million acres of land in West Virginia, Virginia, Kentucky, Tennessee, and South Carolina. Almost half its land is natural hardwood forest, and much of the rest is pine plantations. In 1999 Westvaco signed an agreement with the Nature Conservancy to jointly conduct a five-year survey of company lands to identify rare, endangered, and unique plant and animal habitats. Westvaco has also entered into management agreements with Ducks Unlimited, the National Wild Turkey Federation and the Ruffed Grouse Society.

Westvaco calls its approach ecosystem-based multiple-use forest management. Company lands are placed in zones that emphasize water

At the McGregor Model Forest in British Columbia, professional foresters work with members of the community to collect data about the forest. Canada was an early leader in establishing model forests to develop, test, and disseminate ideas associated with sustainable forestry. Several other model forests have been established in other provinces and in developing nations. PHOTO COURTESY OF McGREGOR MODEL FOREST ASSOCIATION.

quality, nonforest management, special areas, timber management, habitat diversity, and visual quality. About 20 to 30 percent of all of Westvaco's forestland is in zones dedicated to protecting environmental and aesthetic values.

Sustainable forest management requires continuing inventory and monitoring information. To make decisions that sustain soil, water, wildlife habitat, and timber values, forest managers must have information about five soil types, forest stand ages, wildlife populations, and water quality.

Most of the harvesting is done by clearcutting because in the natural hardwood stands, the most valuable species (oak, black cherry, and yellow poplar) require direct sunlight for regeneration. Clearcutting blocks that average 45 acres allows these species to outcompete the shade-tolerant species. Even-aged management is also used in the pine plantations, which are managed primarily to produce paper. Using selected, fast-growing seedlings, these forests are planted and harvested in as few as 25 years.

Like many forest products companies in the United States, Westvaco Corporation has developed conservation agreements with several not-for-profit environmental organizations to improve wildlife habitat on company lands. Wetlands are particularly important for wading birds like the great blue heron. Westvaco's land is certified under the Sustainable Forestry Initiative (SFI) with AF&PA. PHOTO COURTESY OF WESTVACO CORPORATION.

Like all forest products companies, Westvaco is in business to make a profit. It seeks to achieve a reasonable rate of return on its investments by harvesting and regenerating its forests. Westvaco has harvested 46 million tons of wood from its lands since 1980 and at the same time has increased its standing forest inventory by 50 percent. By cooperating with conservation organizations like the Nature Conservancy, Westvaco has also been able to identify critical habitats and manage them to ensure that the required habitat features are available and that the company can still operate profitably.

Green Labels

Some certification efforts involve a "chain of custody" to ensure that final wood and paper products are made from trees that originated from certified forests and to ensure that milling and manufacturing practices are environmentally benign. Retail products that comply are given a "green label" that certifies both sustainable production and manufacture.

In Canada, Europe, and the United States, some environmental groups have persuaded several large retailers not to sell wood products from endangered forests or environmentally sensitive areas. Prominent examples in the United States are Home Depot, Lowe's, Wickes, and 84 Lumber. As a result, the projected demand for certified products has increased in a very short period of time. In 2000, only a few percent of the lumber sold in the U.S. market was certified. The amount seems likely to increase in the near future.

THE CHALLENGES TO SUSTAINABILITY

E fforts to implement sustainable forest management face important obstacles. Human population growth is the most serious, but others are almost as daunting. In many nations, the political and economic institutions necessary for conserving and managing forests are weak or absent. Increasing demand for forest products is another challenge. Global climate change will likely affect our ability to maintain existing forests and create new ones. The challenges require our understanding.

Population

Many scholars have identified human population growth as the most significant factor in the destruction of natural ecosystems, and indeed, the historical record provides strong evidence for a relationship between population and deforestation. First, human population growth drives forest clearing for food production, and second, wood has been the principal fuel and building material for almost every society for more than 5,000 years. As societies grow, they cut and consume more wood. Conversely, when a society declines, forests tend to regenerate. One exception is in the developed nations, where substitution of petroleum energy for wood fuel and the development of accelerated plant breeding have allowed agricultural production to be concentrated on fewer acres, permitting forestland to regenerate even as populations grow.

In *State of the World's Forests 1999*, the United Nations' Food and Agriculture Organization (FAO) concludes,

> The major causes of change in forest cover in the tropics appear to be expansion of subsistence agriculture in Africa and Asia and large economic development programmes involving resettlement, agriculture and infrastructure in Latin America and Asia.

Similarly, in a 1998 article, Mather et al. report,

> ...approximately half of the variation in extent of deforestation is explained
> in statistical terms by variation in population...Over the span of human
> history, population can thus be viewed as a primary driver of deforestation.

The FAO reports that much of the forest loss in Africa is the result of expanded subsistence agriculture. In Latin America, centrally planned government resettlement schemes, large-scale cattle ranching, and hydroelectric development are associated with forest loss. In Asia, FAO sees both types of forest change in approximately equal proportions. Each of these factors is closely associated with human population growth.

Human population reached the 1 billion mark about 1825. There were 2 billion human beings about 100 years later, in 1925. In 1960, after just another 35 years, the population was 3 billion; in 1975, 4 billion; in the late 1980s, 5 billion; and in 1999, 6 billion. Thus the world's population roughly tripled in the 75 years between 1925 and 2000. Figure 2 presents the U.S. Census Bureau projections for world

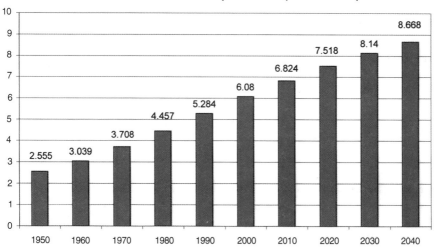

Estimated World Population (in Billions)

Figure 2. Human population more than doubled from less than 3 billion to more than 6 billion between 1950 and 2000. Conservative estimates project a population of nearly 9 billion by 2040. Source: U.S. Census Bureau.

population; it estimates that the human population may reach more than 8 billion by 2040.

We increase by 90 million people each year. The rapid rise occurs because mortality rates have declined while birth rates have remained steady or increased. The annual growth rate is not uniform, however. It is highest in the semiarid and tropical zones of Africa (about 3 percent), the Middle East (2.9 percent), and Central America and Mexico (2.2 percent). In most of the industrialized nations, the annual rate of increase is less than 1 percent: 0.3 percent in Japan, 1 percent in the United States, 0.4 percent in Western Europe as a whole, and –0.1 percent in Portugal.

Human population in the most rapidly growing regions will double in about 23 years if the rate of increase remains constant. Even at a relatively low rate of increase, like 1 percent, population doubles in 70 years. It is the raw power of exponential human population growth

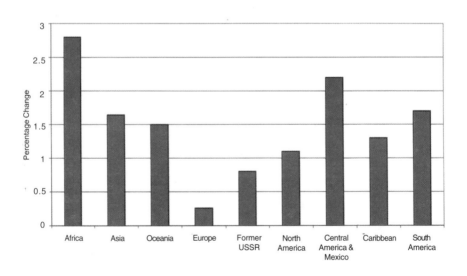

Figure 3. Human population growth rates are greatest in Africa, Central America and Mexico, South America, and Asia. These are also the regions that are losing their forests most rapidly. Ultimately, sustainable forest management requires balancing human population with resource use. Source: FAO.

and the relatively slow rate of growth of many forests that challenge global forest sustainability.

The industrialized countries have gone through something demographers call the "demographic transition." In nonindustrialized countries both birth rates and death rates are high. As public health and sanitation programs increase, the death rate decreases, but the birth rate remains high and the population grows rapidly. As education and living standards rise and families make the transition from agricultural work to industrial work, the costs of bearing and raising large families increase. Birth rates begin to decline and gradually approach death rates.

The world experienced its highest population growth rate in the late 1960s. Birth rates in many countries have since declined. In much of the developing world, public health and nutrition programs are beginning to reduce infant mortality, but the transition to industrialization is slow. As a result, even conservative and moderate

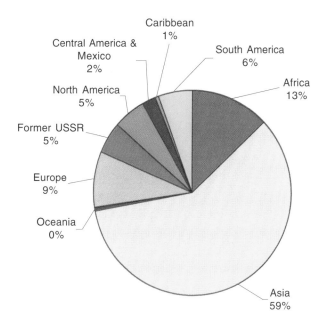

Figure 4. The combined human population of Asia and Africa is almost three-quarters the total estimate of 5.7 billion. SOURCE: FAO.

projections place human population at 10 billion or more by 2050. The majority of the population growth will continue to take place in the developing nations of the tropics. Figure 4 displays the distribution of human population by region in 1995.

The question for the purposes of forest sustainability is what effect a near doubling of these nations' populations will have on the existing forest in the next 50 years.

Political and Economic Institutions

Population growth does not have to mean deforestation. Europe and North America are places where forest area has increased as population has grown. But in the absence of institutions for managing the relationship between humans and the environment, population growth is highly correlated with adverse effects. If governmental controls, markets, and property mechanisms are lacking, deforestation seems inevitable.

Many of the nations that are experiencing the most rapid rates of both deforestation and population growth are still recovering from problems associated with economic exploitation by colonial powers from about 1600 to as late as 1960. Typically, these countries had traditional community rules for allocating property and authority. Common property regimes, in which members of clans or communities determined resource allocation, were widespread.

One of the first acts of the colonial powers was to undermine the authority of local leaders in order to secure their own power. As a result, the old institutions that controlled resource allocation and common property systems broke down. The colonial governments controlled wealth by distributing the power to use the land and its resources to Europeans and their colonial allies. Natural resources that were previously managed by clans, families, and tribes were no longer strictly allocated, and rapid resource depletion was often the result. This should not suggest that resource degradation did not occur in indigenous cultures; nevertheless, changing the economic and political rules often had negative consequences for the forest.

By the 1800s some colonial powers saw that resource depletion was preventing economic development. The typical response was to establish forest reserves that excluded local people and put the land under management of European-trained foresters. The lands outside the

Throughout much of the developing world, new emphasis is being placed on making sure that local citizens make decisions about how local forests are managed and that they receive the benefits of caring for and growing new forests. PHOTO COURTESY OF STEVEN ANDERSON.

reserves were now unmanaged, and once those resources were depleted, people sought to steal resources from the reserves. In Kenya, for example, conflict over use of the protected forest reserves was one factor that sparked rebellion against the British. In almost all colonies, land and wealth became concentrated in the hands of a few individuals or tribes. Lacking secure title to the land, people had no incentive to make the long-term commitments and investments necessary for conservation and sustainability.

In many developing countries, institutional challenges, such as poorly developed or nonexistent property systems, governmental regulations, and markets for forest products, remain the barriers to sustainability. In much of the developing world, community forestry or "social forestry" now encourages individuals and communities to plant and tend forests, making certain that the benefits of forests are obtained by those who care for them. In the recent movement to establish criteria and indicators of sustainability, these institutional issues play a critical role.

In much of the world, wood still provides most of the fuel for cooking, washing and heating. Near Hyderabad, India, this merchant sells firewood gathered from a nearby forest to local residents. PHOTO COURTESY OF STEVEN ANDERSON.

Forest Products Consumption

In the past, human societies were literally built on wood, and global use of wood continues to grow. Consumption expanded by one-third from 1970 to 1995, reaching 3,350 million cubic meters; 63 percent of this volume was used for fuelwood.

Between 1970 and 1994, fuelwood and charcoal consumption increased by 60 percent. In the developing countries, wood fuels accounted for 91 percent of the wood harvested in Africa, 82 percent in Asia, and 70 percent in Latin America. Two of every five people worldwide rely on fuelwood or charcoal for cooking and heating. The FAO estimates that 100 million people already face a fuelwood "famine," and that by 2000, 3 billion humans will rely on wood energy.

Those statistics might make the reader envision hordes of desperate firewood gatherers destroying the forests in pursuit of fuel. But does use of wood for energy lead directly to deforestation? Because fuelwood is not often traded in the formal economy, many governments do not keep track of it. Definitions for fuelwood and methods for its estimation vary. Whatever the deficiencies in the data, after much concern about a fuelwood crisis in the 1970s and 1980s, in 1997 the FAO asserted,

> …in most cases, fuelwood collection is not a primary cause of defor-estation…it is now clear that fuelwood production and harvesting systems can be and often are, sustainable.

The balance of the world's wood consumption falls in a category the FAO calls "industrial roundwood," for which statistics are reported in national economic reports. This includes saw logs, veneer logs, pulpwood, and in some cases chips, particles, and wood residues. The FAO predicts industrial roundwood production and consumption will increase from about 1,490 million cubic meters in 1997 to 1,870 million cubic meters by 2010. The largest increases in production and consumption will come in Asia, with slower growth in Africa and South America. Europe, Asia, and North and Central America will account for about 85 percent of production and more than 90 percent of con-sumption by 2010. Asia will remain the world's only net importing region for industrial roundwood.

The FAO concludes that the capacity of the forest and other sources of fiber will be sufficient to meet demand for the foreseeable future. However, some regions will see shortages in some categories (e.g., large logs). Demand for high-quality saw logs will approach or even exceed the production capacity of forests and plantations in Africa, Southeast Asia, and the Pacific Islands.

Much of the optimism associated with the availability of industrial roundwood is based on improvements in the efficiency of wood processing. For example, in 1970, 1 tonne of paper and paperboard typically contained 80 percent wood pulp; today, according to the FAO, the figure is approaching 50 percent. More efficient wood products technologies are transforming the industry in the developed countries. Engineered wood products and wood-based panels are increasingly substituted for traditional products. Forest industries are generally much more efficient in using raw materials than they were 30 years ago, and recycling and residue recovery have had an effect. As a result, the use of industrial roundwood grew only 15 percent between 1970 and 1994.

The challenge lies in extending this efficiency to developing nations as they increase their use of wood as an industrial product. The rate at which these new technologies are adopted is one factor that will determine the future of forest sustainability.

The FAO concludes that existing forest resources are adequate for industrial needs for the foreseeable future. Based on the FAO data, it does not appear that current levels of consumption of industrial forest products are a primary influence on the extent of the world's forest cover. In a 2000 report, however, the World Bank was less sanguine:

> The poor have been less a source of deforestation in the forest-rich countries than [previously] assumed. A much stronger factor has been the growing domestic demand for fuelwood in industries, for timber in housing construction and for tropical forest products in international markets.

The World Bank concludes that without management of international demand for tropical timber and adequate compensation for countries that incur costs for achieving global environmental objectives, deforestation will continue.

Plantation Forestry and Intensive Forest Management

Planting forests for wood consumption goes back to the Ptolemaic dynasty of Egypt (323–30 B.C.). To avoid reliance on imported timber, the Egyptians embarked on a massive tree-planting program, built government nurseries, and regulated harvest practices on private and public lands.

In modern times, the area of forest plantations increased rapidly after World War II in the developed countries and after 1960 in the developing countries. In the United States, planted pine plantations became common in the South during the 1930s.

Plantation forests are most often used to meet demand for industrial wood, but they also produce wood fuels, fodder, and other forest products. Asia is, by far, the leader in plantation forests. China, India, Indonesia, Viet Nam, and the Republic of Korea all have more than 1 million hectares of forest plantations of varying quality. Plantations can supply significant proportions of the demand for industrial roundwood in a small space. For example, Argentina and Brazil meet about 60 percent of their industrial roundwood needs from plantations that occupy about 2 percent of their forest area. Chile and New Zealand meet more than 90 percent of their industrial roundwood demand from plantations that occupy 16 and 17 percent of their total forest cover, respectively.

There are tradeoffs involved in plantations, however. Plantations that replace "natural" forest may produce significantly different plant and wildlife communities, but where overharvesting is a cause of deforestation or forest degradation, plantations may help protect the remaining natural forest. Increased forest plantation development has been a source of conflict in some communities. Controversies over the species planted and the right to use the fodder, forage, and wood from the plantation are often a problem where property rights and responsibilities are not clearly articulated.

Meeting the increased demand for wood products while protecting the ability of the earth's forests to provide wildlife habitat, clean water, and a host of other environmental services is one of the principal challenges for professional foresters. In a global, market-based economy,

Converting old cropland to pine plantations became common in the South during the 1930s. Until then few landowners had thought of trees as crops. The potential of forest fires made tree farming risky until state forestry organizations were created for fire control. This plantation was established in 1936 in Lee County, Alabama. AMERICAN FOREST INSTITUTE, COURTESY OF FOREST HISTORY SOCIETY.

In many places—here, the Willamette National Forest—foresters have used clearcutting to harvest and subsequently regenerate forests. The practice is controversial because it changes the qualities and distribution of wildlife habitats and requires extensive road systems. PHOTO COURTESY OF USDA FOREST SERVICE.

human demands for wood fuel, paper, and lumber will be met if there is a profit to be made.

An important issue in sustainable forest management is the role of intensive forest management in meeting market demands while protecting the remaining natural forests that are environmentally sensitive. By intensively managing some forests for fiber and fuel production and managing other forests to maximize biodiversity conservation, wildlife habitat, or watershed values, sustainable forest management attempts to achieve the combination of forest uses that maintains the forest as well as the human community that depends on it.

It seems likely that some of the world's demand for fuel and fiber will be met by new crops and improved recycling technologies. But intensive forest management of some forests will be required to conserve other, environmentally sensitive forests. Like modern agriculture, intensive forest management may rely on controversial practices like clearcutting, the use of chemical fertilizers and pesticides, and the use of genetically modified species. It may also mean applying improved techniques of managing naturally regenerated forests.

Some forest products companies, governments, and individual landowners practice intensive forest management in "natural" forests. The highly productive private forests in the coastal regions of the Pacific Northwest and in the Southeast are examples of this type of forestry. It seems reasonable to expect that the use of intensive forest management will increase in the coming decades. Without wisely applying the science and technology that are available to us, we diminish our chances of protecting the environmentally sensitive forests that remain.

Climate Change

Humans have been and remain the most potent force in changing the extent and nature of the world's forests during the last 10,000 years. But they are not the only influence. The earth's climate is dynamic, as are the geologic processes that have determined the distribution of the world's biota.

Because of repeated cycles of warming and cooling, the distribution of the world's forests and arable lands fluctuated long before the beginnings of human history. A cooling period at the beginning of the Cenozoic period (65 million years ago) led to expansion of temperate forests over the Northern Hemisphere. These forests then retreated with the advancing ice sheets of the Pleistocene. The last glaciers receded about 10,000 years ago.

Long-term climate change data come from ice cores. The oldest records are from Antarctica and Greenland and extend about 250,000 years. At the end of the Wisconsin glacial period (about 18,000 years ago), global temperatures reached their coldest point in the past 100,000 years. The warming trend since then has fluctuated several times. It reached a maximum (about 1 degree C warmer than now) about 5,000 years ago; then temperatures dropped slowly for several thousand years before rising briefly in the Middle Ages and falling again. A general warming trend has dominated the atmosphere since 1800, and in the past 100 years, the average temperature has increased about 0.5 degree C.

Carbon dioxide concentrations have increased rapidly since the Industrial Revolution from a little more than 270 parts per million (ppm) in 1730 to more than 360 ppm in the late 1990s. In conjunction with methane, chlorofluorocarbons, tropospheric ozone, and nitrous oxide, these "greenhouse gases" are increasingly believed responsible for increases in global temperatures. There is much speculation about the ability of forests to absorb atmospheric carbon. A global increase in carbon could stimulate photosynthesis and growth in plants. If some of it is stored in plants, forests could act as a sink, buffering global warming. Yet, increased photosynthesis might lead to increased plant respiration, which returns carbon dioxide to the atmosphere. Change in wildfire patterns might also release more carbon.

Global climate change is more likely to influence the distribution of forests and agriculture by changing the hydrologic cycle and soil moisture, which could affect the distribution of plants and animals perceptibly in the course of a human lifetime. There is growing consensus among scientists about the mechanisms and the effects of human-induced global climate change. Some governments have been reluctant to enact significant policy changes because they fear economic dislocations. Recently, disagreements about how to count actions like planting new forests appear to have delayed international agreements.

Defining the Forest

If sustainability means meeting the needs of current and future generations, we must also know how much forest exists to meet those needs and accurately assess the rate of change. This challenge begins by estimating how much forest existed in the past. Did the changes occur because of global climate change or did they occur to promote another land use, like agriculture? Are the estimates of the current forest area based on surveys of knowledgeable officials, on-the-ground measurements, or electronic images from aircraft and satellites?

But then what is a forest? Should plantations of trees be considered forests? If a forest is cleared for agriculture but regrows into a new forest in ten or 20 years, is that sustainability or deforestation? Many of the forests around the world have been cut down and converted to agriculture and regrown several times in the past thousand years. Does this indicate some measure of sustainability?

Semitropical and semiarid regions are sometimes covered with woodlands. Although not classified as forests in many inventories, they provide important wildlife habitat, fuelwood, and forage for livestock. PHOTO COURTESY OF STEVEN ANDERSON.

Answers to those questions depend on one's values; they are not issues of science. Nevertheless, science can help us understand how and why forests grow and change. Estimates of global forest loss presume a shared definition of *forest*. According to the *Dictionary of Forestry,* a forest

> is an ecosystem characterized by a more or less dense and extensive tree cover, often consisting of stands varying in characteristics such as species composition, structure, age class, and associated processes, and commonly including meadows, streams, fish, and wildlife.

The Food and Agriculture Organization of the United Nations has different definitions for *forest*. In the developing countries (primarily in the Southern Hemisphere), a forest is an "ecosystem with a minimum of 10 percent crown cover (the area of ground shaded by overhead sun) of trees and/or bamboo generally associated with wild flora, fauna and natural soil conditions, and not subject to agricultural practices." But in the developed countries (primarily in the Northern Hemisphere), the crown cover must be more than about 20 percent. According to these definitions, if the vegetative cover in a forest in Idaho goes below 20 percent, the forest is "deforested." If the same thing happens to a forest in Madagascar, deforestation would not have occurred. Further complicating "forest" comparisons is that FAO distinguishes between a "forest" and "woodland"—a combination of grassland and trees. In much of the world's semiarid regions, vast areas are covered by grasses, shrubs, and occasional trees. These ecosystems provide valuable wildlife habitat and watershed protection, but they are not counted as forests in most assessments.

To understand the change in the extent of the world's forests, one must first estimate the original area of the forest—the forest cover—and then track the changes at regular intervals. Some nations began estimating the extent of timber for naval reserves hundreds of years ago when it became apparent that forests were important to national defense. It was only in the 20th century, when we recognized forests as

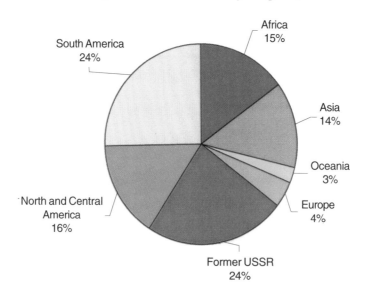

Figure 5. Nearly half the world's forests are in South America and the former Soviet Union. SOURCE: FAO.

Percentage of Tropical Forest Area by Region, 1995

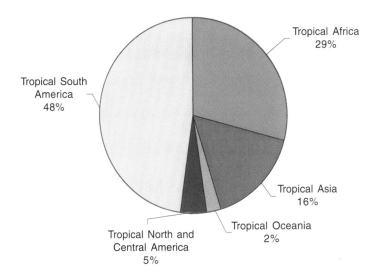

Figure 6. Tropical forests have become the focus of much concern because of their biological diversity and because they are being converted to other uses at a rapid rate. Most of the world's remaining tropical forests are in South America and Africa. SOURCE: FAO.

producers of transnational values, that concern shifted to a global scale. Assessment, however, has largely remained the responsibility of individual nations. Many countries do not inventory their forests at regular intervals, and assessments of forest changes are often the result of educated guesses. We are just beginning to develop universal systems to assess the extent of the world's forests and their attributes, such as biological diversity and watershed protection.

With many nations and cultural views of a forest come many definitions. Because there is no single definition of *forest*, there is no single definition of *deforestation*. That is just one of the problems associated with estimating the extent of the forest and the way that forest cover is changing. If one is interested in sustaining the other attributes of the forest (biodiversity, or timber, or ability to capture and hold carbon, for example), the complexity of assessment multiplies.

The Challenge of Global Assessment

As concern for global approaches to forest assessment grew in the 1970s, it became apparent that existing tools for measuring the world's forests were inadequate. The problem had been recognized before, however. In 1923, in the introduction to the first widely circulated global timber assessment, Gifford Pinchot wrote that President Theodore Roosevelt wanted to convene a worldwide natural resources assessment conference out of concern about a possible timber famine. The 1923 estimate of forest cover, based on aggregated national reports, was a little more than 3 billion hectares.

The FAO began attempting to estimate global forest cover in the 1940s. But even in the early 1990s an important textbook on inter-national forestry reported that the extent, condition, and trend of global forest cover could not be described with precision. According to a 1992 compilation by Laarman and Sedjo, estimates of global forest cover between 1960 and 1990 varied from some 2 billion hectares to 6 billion hectares.

Since the late 1990s several organizations have been trying to assess global forest cover. They include the FAO, the International Geosphere-Biosphere Programme (a unit of the European Commission), and the World Conservation Monitoring Center. Each approaches the problem somewhat differently. Some rely on "bottom-up" national surveys, and others take "top-down" approaches based on satellite imagery. Several of the assessments concentrate on the tropics; others are wider in scope. Unfortunately, these assessments use different definitions, sources, measures, and methods for classifying vegetation, and thus their estimates of forest cover are not directly comparable.

The FAO's Forest Resource Assessment 2000 is gathering information on several attributes in addition to timber information, including nonwood forest products, naturalness, biological diversity, protected areas, carbon sequestration, forest condition, and socioeconomic functions. This is an ambitious goal indeed. Knowledge of the distribution and status of tropical animal and plant species is very limited. Rough approximations of species richness are being developed and "biodiversity hotspots" have been identified, but mapping and classification are complex tasks, and the job of collecting, organizing, and monitoring environmental information in developing countries can be immense.

In the 1970s one commentator noted that science had produced better assessments of the moon's craters than of the world's forests. That situation is slowly being remedied.

The Global Forest

Defining the forest and assessing its qualities are essential to resolving questions associated with sustainability, and global assessments are needed so that the efforts of governments and other organizations can be evaluated. Although many of the data from the FAO are the result of national forest inventories taken only once in the past few decades, and accuracy is questionable, they give some idea of the extent and condition of the world's forests.

The FAO estimated in 1990 that forests cover about 3.4 billion hectares (about 13.3 million square miles), or about 26 percent of the earth's total land area. If one uses the broader classification of forest and woodland, the total is estimated at 5.1 billion hectares (about 19.7 million square miles), or 39 percent of the earth's total land area. The FAO also reports recent changes in forest cover, but to better understand the changes through human history, we have to examine other evidence.

Several geographers have attempted to estimate the forest cover of the earth before human population growth became such an important factor. Any single-number estimate should be used with caution because

Change in the World's Wooded and Forested Land Area

	Hectares	Square kilometers	Square miles
Original wooded land	7,770,000,000	77,700,000	30,000,747
Wooded land in 1990	5,100,000,000	51,000,000	19,691,610
Forested land in 1990	3,454,000,000	34,540,000	13,336,239
Annual net loss of wooded and forested land, 1980–1995	12,000,000	120,000	46,333

Table 2. Three units of measure are often used to discuss changes in forest area. For easier comparison, this table presents estimates of original wooded land, wooded land in 1990, forested land in 1990, and annual net loss of wooded and forested land in hectares, square kilometers, and square miles. SOURCES: MATHER ET AL. (1998), FAO.

of changes in climate, but after looking at potential natural vegetation, Mather et al. produced an estimate of 77.7 million square kilometers of original forest cover. Accepting their work and the FAO's assessment of total forest and woodland (51 million square kilometers) as rough approximations, the earth has lost a little more than a third to half of its total forest cover during the past few thousand years of increasing human dominance.

One of the reasons for increasing concern about sustainable forestry is the rapid rate at which forests are disappearing. The FAO estimates that between 1980 and 1995, the earth lost about 180 million hectares of forest—an area equivalent to the total area of Mexico.

The greatest losses are in the tropics and the world's developing nations. In the first ten years of the period (1980–1990), the annual loss in the region was about 15.5 million hectares. This rate slowed to about 13.7 million hectares between 1990 and 1995—a loss of less

Regional Change in Forest Area, 1990–1995

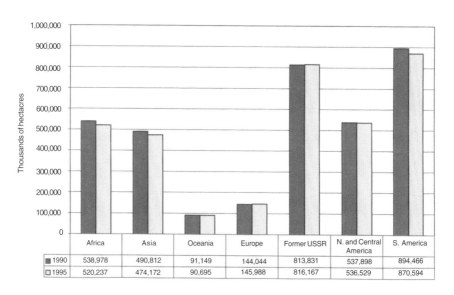

	Africa	Asia	Oceania	Europe	Former USSR	N. and Central America	S. America
■ 1990	538,978	490,812	91,149	144,044	813,831	537,898	894,466
□ 1995	520,237	474,172	90,695	145,988	816,167	536,529	870,594

Figure 7. In Africa, Asia, and South America there were significant losses in forest area between 1990 and 1995. Although the differences in the bars are small, the change is important because it represents a very short time period. SOURCE: FAO.

than 1 percent per year. Although the percentage rate seems small, the effect is serious. In total, the tropics and developing nations lost about 200 million hectares of forest between 1980 and 1995. Clearly, the first challenge to sustainable forestry is conserving the remaining tropical forests.

The total loss and the rate of loss are alarming. The FAO predicts that the food demands of the growing human populations concentrated in the developing countries will cause additional forest clearing. Food demands are likely to increase at a rate of 1.8 percent per year through 2010, requiring an additional 90 million hectares for food production (mostly in sub-Saharan Africa and Latin America). Much of that area will be converted from existing forest.

A small portion of global forest loss has been offset by gains in forest area in the developed countries. According to FAO, developed nations (mostly in North America and Europe) gained 20 million hectares of

Change in Tropical Forest Area, 1990–1995

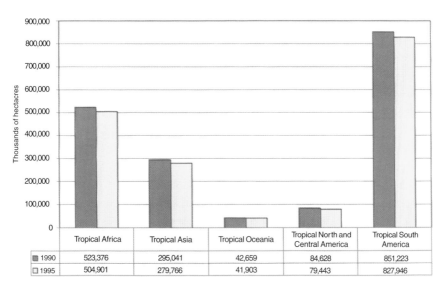

	Tropical Africa	Tropical Asia	Tropical Oceania	Tropical North and Central America	Tropical South America
1990	523,376	295,041	42,659	84,628	851,223
1995	504,901	279,766	41,903	79,443	827,946

Figure 8. Forest losses are greater in the tropics than any other region because of rapid human population growth there. The overall losses are greatest in South America, Africa, and Asia. Central America and Africa had the greatest rates of tropical deforestation during the period. SOURCE: FAO.

forest from 1980 to 1995. This gain more than offset the losses in the developed countries due to urbanization and infrastructure development.

The U.S. Forest

The United States saw major deforestation and agricultural expansion between 1850 and 1910, when American farmers cleared more than 190 million acres of forest—about the same area as our current national forest system. Estimates of the original U.S. forest cover in 1600 are a little more than 1 billion acres. About 30 percent of that (307 million acres) was eventually cleared, mostly for agriculture. The forested area of the United States stabilized and even grew slightly during the late 20th century. Today, about 737 million acres—two-thirds of the original forest or 33 percent of the total land area—is covered by forest.

Those data describe changes at the national level, but there is much that they do not reveal. Many of the forests of New England and the South have regrown following abandonment of agriculture. The old stone fencerows seen in the forests of New York and New England reveal that the region has considerably more forest now than it did in 1850, when the Midwest captured a competitive advantage in food production. Much of the current loss of forest cover in the United States is localized near urban areas and transportation corridors. States like Michigan and Minnesota are losing forestland to urban and suburban growth, and in many places citizens are alarmed about deforestation and forest fragmentation at the edges of cities. Although the United States on the whole has gained forestland, the loss of forest in some areas and the change in the nature of the remaining forest concern many citizens.

The U.S. Department of Agriculture tracks changes in privately owned land use in the United States through the National Resources Inventory (NRI). Private land is most subject to change in land use. Since 1982 there has been a net gain of about 800,000 acres of privately owned forest in the United States. From 1982 to 1997, the United States gained about 19 million acres of forest from pasture and cropland and lost about 18 million acres of forest, mostly to urban development. The change in forestland is much smaller than the changes in cropland

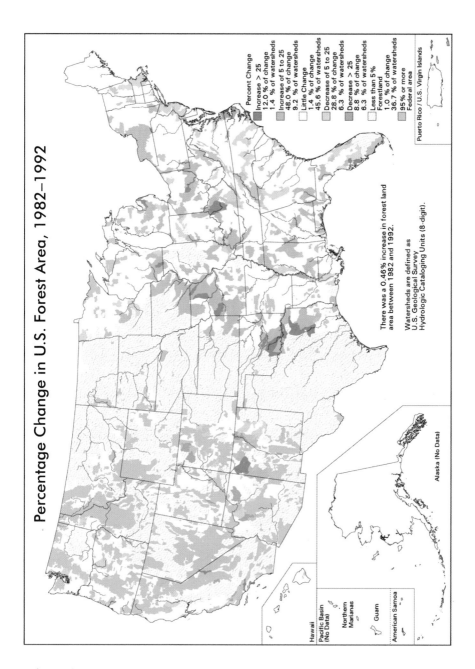

Figure 9. Changes in forest area within the United States vary from region to region. Forest losses are notable from southern New England to Washington, D.C., in Florida, and along some major river corridors. Gains in forest area are apparent in the Great Plains, Midwest, and upstate New York. SOURCE: USDA NRCS.

and developed land. The national data indicate a net increase in forestland, but many areas that surround major cities are losing outlying forests and farmland.

It is difficult to overemphasize the importance of both temporal and spatial scale in discussing sustainability. Many suburban residents are alarmed as development swallows farmland—land that was forest in 1600—but it is difficult to see the gradual expansion of forests in the United States since they reached their historic minimum in the 1920s. In fact, U.S. forest area has grown slightly, even as the population and the demand for forest products have grown.

Changes in Forest Composition

The forest itself is changing. Left alone over long periods, forests often change in composition and extent as a response to disturbance like fire and hurricanes. But for the last 400 years in North America, most of the forest change has been created by humans. Broad-leaved, hardwood forests are replaced by coniferous softwoods in some regions; elsewhere the change is from softwoods to hardwoods. One important way that many forests are changing is called fragmentation, which occurs when large areas of forest become small patches interspersed with other land uses. An airplane flight over the Midwest is the easiest way to see forest fragmentation. As forest areas become smaller, the number and kinds of wildlife they contain can change dramatically as species that require large amounts of interior forest habitat are displaced by species that thrive at the forest's edge.

Many "natural" forests have been converted to managed forests. In a managed forest, natural processes are manipulated to produce a mix of values and products that match human demands. In some cases, plantation forestry has replaced the original forest with genetically modified species and varieties selected for rapid growth and forms better adapted for timber production. For example, the Forest Service projects that plantation forests in the South will increase from 36.1 million to 45.2 million acres between 2000 and 2040. Many of these plantation forests will replace land that is currently in agriculture.

Similar transformations are occurring in Europe. The WorldWide Fund for Nature reported in 1994 that less than 1 percent of the forest of western and northern Europe retained much of its natural character.

However, the area of forest cover is increasing across the continent, and in many places forest area is increasing at the same time the human population is growing. Denmark, France, Italy, Portugal, and Switzerland have all dramatically increased the percentage of forest cover since reaching minima in the 1800s or early 1900s.

The condition and health of the forest are also changing. Here, data are even more elusive, but very large forest fires and locally important outbreaks of diseases and insects are changing the world's forests in important ways. The FAO estimates that roughly 2 million hectares of forest in the temperate, subtemperate, and humid tropical regions burned each year during the 1980s. The estimate does not include the dry tropical zone. A forest fire does not necessarily lead to deforestation, but without better monitoring, especially in semitropical and tropical forests, the effects of fire remain unknown.

Many people in the developed and developing countries are rightfully concerned about deforestation in the tropics, but this is not a new phenomenon. The United States, Canada, and Europe underwent similarly alarming changes in forest cover during their history. There are important differences in the nature of the soils and species diversity that distinguish tropical from temperate forests, but the processes of deforestation are essentially the same.

Harvesting in and of itself does not necessarily cause deforestation. For example, forest products harvesting expanded considerably in the United States during the 20th century. At the same time the area and volume of the nation's forests also expanded. But the question is still hotly debated in much of the developing world. Clearing for agriculture remains the primary cause of deforestation in the tropics, and exploitative logging is a concern in some temperate and boreal forests.

Poor harvest practices can and do damage soil and watershed values. Overharvesting of commercially valuable species remains commonplace. Roads built to move timber to markets often open previously inaccessible forests to commercial exploitation and clearing for agriculture. Developing the institutions that allow proper forest management takes time, but the demands of rapidly expanding populations make it more difficult for governments to establish the institutions that protect forests.

In many ways, the forest today is different from that of 1600 in North America or 1000 in Europe and Asia. How one perceives those changes is closely related to the problem of assessing forest sustainability. People who seek fuel and fodder or lumber and paper may applaud the increase in managed forests and the rapid growth of trees in plantations. People primarily concerned with biological diversity or nurturing indigenous cultures may want to sustain a very different kind of forest.

We clearly aspire to sustainability. A recent report by the USDA Committee of Scientists (1999) suggests that sustainability is a "guiding star." As society begins to follow the star of sustainability, we struggle with differences in definitions and the limitations of our knowledge.

CONCLUDING THOUGHT

For approximately the first 1.5 million years of human existence, the environment had shaped the animal. With the beginning of agriculture, that relationship changed. In the last 10,000 years humans have increasingly shaped the environment. The relationship remains dynamic, but in recent centuries, humans have arguably had a much greater influence on the environment than the environment has had on us.

For the many billions who live in the developing nations, the day-to-day needs of survival remain paramount. But in the developed nations, more and more of us seek to address problems at large geographic scales over long periods of time.

Sustaining the world's forests is a critical goal for our society, and the importance of sustainable forestry cannot be overstated. In the last 15 years, significant steps have been taken to define the concept and to measure progress toward its goals. That almost all nations, industries, and interest groups involved in forestry now agree on a set of goals is a significant achievement, but that in itself will not result in appropriate stewardship. The concept of sustainability remains amorphous, and some consider it more a philosophy than a set of management practices or observable condition of the forest.

To be successful forest managers, forest users and forest policy makers will have to make rapid progress on difficult challenges. The longer it takes to implement the changes, the more forest will be lost.

The principal challenges to sustaining the forest in the developing world are population growth and large-scale resettlement and economic development schemes that convert many of the world's tropical forests to fields for producing food. Managing population growth must be the first priority of those concerned with protecting and conserving the earth's forests and the creatures that live within them.

The nature of the earth's forests will continue to change and the total extent of those forests seems likely to continue to decline in the next 50 years. In nations where political, market, and property institutions are strong and citizens consider the legacy they will leave, forests may thrive. In nations with weak institutions, and where just finding enough fuelwood to cook tomorrow's food is a principal concern, the outlook is rather grim. The gulf between these two worldviews is enormous.

The scientific knowledge we have gained in the past century about ecosystems, genetics, biological diversity, global nutrient cycles, and energy is genuinely astounding. In some ways it seems odd that despite this new knowledge we are still confronted with the same problems of sustainable agriculture and sustainable forestry that confounded people thousands of years ago in Sumeria, Greece, India, China, and Meso-america. That might lead one to conclude that the problems of sustainability cannot be solved by science alone. Science is crucial to sustainability, and implementing sustainable forestry will require new scientific knowledge and new cost-effective technologies. But science is only one part of the answer.

In a new book that revisits their findings in *Limits to Growth,* Meadows et al. conclude that the use of some natural resources has already exceeded the physical limits of sustainability. Decline is not inevitable, they argue: They believe that achieving sustainability is technically and economically possible. But it will require limiting the growth of material consumption and human population and dramatically increasing the efficient use of energy and resources. In other words, achieving sustain-ability is possible only if we choose it as a goal and change our behavior accordingly.

Over the past 3,000 years our concerns have widened from local to global and from tomorrow to several generations from now. But while our concerns have evolved, our ability to shape human behavior has changed little. In his 1968 essay *The Tragedy of the Commons*, Garret Harden described a deceptively simple solution to the problem that he posed. He suggested that "mutual coercion, mutually agreed upon" was the answer to overuse of unowned natural resources. Until communities and governments organize and decide to make the choices that will sustain their forests through mutual consent, there is little hope that sustainability can be achieved.

In the pursuit of sustainability we must simultaneously control our population, feed our hungry, retain our forests, and leave happy choices for future generations. Like the search for the Holy Grail, the prize seems likely to elude us, but the quest is essential.

Suggested Reading

Bartlett, A.A. 1997–98 (1994). Reflections on Sustainability, Population, Growth and the Environment—Revisited. *Renewable Resources Journal*. Winter, 1997–98:6–23.

Bromley, D.W. 1992. Property Rights as Authority Systems: The Role of Rules in Resource Management. *In* Nemetz, P. (ed.). *Emerging Issues in Forest Policy*. Vancouver: UBC Press.

Bruntland, G.H. 1987. *Our Common Future: World Commission on Environment and Development*. Oxford University Press.

Caldwell, L.K. 1984. Political Aspects of Ecologically Sustainable Development. *Environmental Conservation* 11(4):299–308.

Clawson, M. 1979. Forests in the Long Sweep of American Forestry. *Science* 204, 15 June 1979: 1168–74.

Cohen, J.E. 1995. *How Many People Can the Earth Support?* New York: W.W. Norton.

Edmonds, R.L. 1994. *Patterns of China's Lost Harmony*. London: Routledge.

Evelyn, J. 1664. *Sylva; or a Discourse on Forest Trees and the Propagation of Timber in His Majestie's Dominion*.

Food and Agriculture Organization of the United Nations. 1997, 1999. *State of the World's Forests: 1997*, and *State of the World's Forests: 1999*.

Frazer, J.G. 1929 (1890). *The Golden Bough: A Study in Magic and Religion*. New York: MacMillan.

Gardner-Outlaw, T., and R. Engelman. 1999. *Forest Futures: Population, Consumption and Wood Resources*. Washington, DC: Population Action International.

Glacken, C.J. 1967. *Traces on the Rhodian Shore: Nature and Culture in Western Thought from Ancient Times to the End of the Eighteenth Century*. Berkeley: University of California Press.

Helms, J.A. (ed.). 1998. *The Dictionary of Forestry*. Bethesda, MD: Society of American Foresters.

Hughes, J.D. 1984. Sacred Groves: The Gods, Forest Protection and Sustained Yield in the Ancient World. *In* Steen, H.K. (ed.). *History of Sustained Yield*

Forestry—A Symposium. Santa Cruz, CA: IUFRO Forest History Group. Forest History Society.

———. 1994. *Pan's Travail: Environmental Problems of the Ancient Greeks and Romans*. Baltimore: Johns Hopkins University Press.

Intergovernmental Panel on Climate Change. 2000. Summary for Policymakers: Land Use, Land-Use Change and Forestry.

Laarman, J.G., and R.A. Sedjo. 1992. *Global Forests: Issues for Six Billion People*. New York: McGraw-Hill.

Leopold, A. 1949. *A Sand County Almanac and Sketches Here and There*. 1987 reprint, New York: Oxford University Press.

MacCleery, D.W. 1993. *American Forests: A History of Resiliency and Recovery*. Durham, NC: Forest History Society.

Marsh, G.P. 1874. *The Earth as Modified by Human Action*. New York: Scribner, Armstrong & Co.

Mather, A.S., C.L. Needle, and J. Fairbairn. 1998. The Human Drivers of Global Land Cover Change: The Case of Forests. *Hydrological Processes* 12:1983–94.

Meadows, D.H., D.L. Meadows, J. Randers, and W.W. Behrens. 1972. *Limits to Growth: A Report for the Club of Rome's Project on the Predicament of Mankind*. New York: Universe Books.

Nash, R. 1967. *Wilderness and the American Mind*. New Haven: Yale University Press.

Natural Resources Conservation Service. 1997. Natural Resources Inventory. Washington, DC: Department of Agriculture. *www.usda.gov/nrcs*.

Perlin, J. 1989. *A Forest Journey: The Role of Wood in the Development of Civilization*. New York: W.W. Norton.

Ponting, C. 1991. *A Green History of the World. The Environment and the Collapse of Great Civilizations*. New York: Penguin Books.

Romm, J. 1994. Sustainable Forests and Sustainable Forestry. *Journal of Forestry* 92(7):35–39.

Stegner, W. 1953. *Beyond the Hundredth Meridian: John Wesley Powell and the Second Opening of the West*. Lincoln: University of Nebraska Press.

Tsoumis, G. 1964. Forestry in Greece. *Yale Forest School News*. 52(3)35–37.

White, L. Jr. 1967. The Historical Roots of Our Ecologic Crisis. *Science* 155, 10 March 1967: 1203–07.

Wiersum, K.F. 1995. 200 Years of Sustainability in Forestry: Lessons from History. *Environmental Management* 19(3):321–29.

Williams, M. 1989. *Americans and Their Forests*. A Historical Geography. Cambridge, UK: Cambridge University Press.

World Bank. 2000. *A Review of the World Bank's 1991 Forest Strategy and Its Implementation*. Washington, DC: World Bank Operations Evaluation Department.

Zon, R., and W.N. Sparhawk. 1923. *The Forest Resources of the World*, 2 vols. New York: McGraw-Hill.

APPENDIX

Conversions for Units of Measure

	Acres	Hectares	Square miles	Square kilometers
Acre	1	2.47	0.0015625	0.0247096
Hectare	0.4047	1	0.0006325	0.01
Square mile	640	1580.8	1	0.38611
Square kilometer	40.47	100	2.5889	1

About the Author

Donald W. Floyd is a professor of forestry and public administration at the State University of New York's College of Environmental Science and Forestry and Syracuse University's Maxwell School of Citizenship and Public Affairs. He has a bachelor's degree from Humboldt State University in Arcata, California; a master's degree from the University of Wisconsin–Madison; and a Ph.D. in renewable natural resources from the University of Arizona in Tucson.

Floyd worked as a seasonal park ranger and forestry technician for the National Park Service before beginning his career as a county extension agent with Colorado State University. He subsequently worked in extension programs in Oregon and Arizona before completing his doctorate.

From 1988 to 1993 Floyd taught at the School of Natural Resources at Ohio State University, where he became head of the faculty of forestry.

During the 1990s, Floyd was a member and chairman of the Society of American Foresters' Committee on Forest Policy and also served as chairman of the Society's Range Ecology Working Group. As chair of its Task Force on Public Lands Policy, he edited *Forest of Discord* (1999). He was elected a Fellow of the Society in 1999.